CANADIAN Geographic

Quiz Book

CANADIAN Geographic

Quiz Book

BY DOUG MACLEAN
Edited by Ian A. Wright

FITZHENRY & WHITESIDE
MARKHAM, ONTARIO

The Canadian Geographic Quiz Book

© 2000 Fitzhenry & Whiteside

Cover design: John Luckhurst / GDL
Interior design: Kerry Designs

Fitzhenry & Whiteside acknowledges with thanks the support of the Government of Canada through its Book Publishing Industry Development Program in the publication of this title.

Canadian Cataloguing in Publication Data

MacLean, Douglas, 1944–
Canadian geographic quiz book

Includes index
ISBN 1-55041-392-9

1. Canada – Geography – Miscellanea. I. Title

FC75.M322 1999   917.1'002   C99-932125-0
                        F1012.M322 1999

This book is dedicated to two people. The first is my father, John, who died while I was putting this book together. Like all fathers he always took an interest in whatever I did but I know he took special pride in my passion for the Canadian environment. He was a geologist and his love of the earth was passed on to two of his sons, both of whom became geographers. My brother Paul is a geomorphologist who owns an environmental-auditing company and I am a geography teacher. During his illness, Dad kept asking me how many questions I had completed. He would have been pleased to see this effort in print.

The second person is my wife, Mary Lynn, the most supportive person in my life. As a young married couple we travelled through all the nooks and crannies that make up Canada and when our daughters, Amy and Stephanie, arrived they travelled with us too. We started with a new Volkswagen Westfalia van which the kids called Morrison and 27 years later it still ferries us across this wonderful country. Goodness knows there were lots of reasons not to travel in any given summer. But thanks to Mary Lynn, there were few summers we stayed home.
So thank you, Dad and Mary Lynn.

Contents

Acknowledgements

Until I was confronted with the task of making up hundreds of questions on more than forty topics about Canada, I never gave much thought to the variety of resources available. For instance, I was given two great books on Canadian disasters, one on Canadian sea monsters and another on Canadian women Olympians. Unfortunately, most of them were so interesting that many a winter evening was spent reading about Ogopogo or Debbie Brill when I should have been typing at my computer. So thanks to all who gave me suggestions for questions and in the process furthered my education.

Special thank you's go to: Mary Lynn, Amy and Stephanie MacLean, Ron Patterson, Eric Lasota, Ian Griffiths, John Gordon, Courtenay Shrimpton, Steve Lee, Brian Moore, David Morton, Gary Harvey, Jon Klinkhoff, Wilma and Gar Grainger, Ken and Claire McGuire, Susan Rose, Gary Millward, and Chris Dupuis. I must give a special thank you to the librarians in my life, Edith Drummond, Judy Johnston, Karen Michaud, Kathleen Miller, and Maria Varvarikos, whose help over the years has always been appreciated. Now how does one thank the Internet, that amazing source of up-to-date data? The Great Canadian Geography Challenge – held annually with thousands of schools participating – helps ignite a passion for Geography amongst Canadian students. I appreciated being able to use some of their questions on Canada for my book. Similarly, thanks to the people who allowed me to use their material in compiling my questions at that special Canadian publication, *Canadian Geographic*. This magazine should be in the home of every Canadian.

I'd also like to thank Ian Wright for his support, his experience, and especially, for his ideas on topics and his suggestions about organizing my material. Thank heavens for Ian at the other end of desperate e-mail messages. His excellent textbook, *Canada – Exploring New Directions*, gave me many ideas, too.

Finally, I owe a big thank you to my editor, Richard Dionne, for his patience and perseverance. His calm, insistent pressure ensured that questions and answers became as readable and well-organized as a book of this nature can be. With me in Montreal and Richard in Toronto, daily telephone contact became the norm. Richard's enthusiasm for this work carried it off.

In spite of my best efforts, there must be a few mistakes in this book. Those errors are mine and not the fault of the people mentioned above.

Doug MacLean

Introduction

Most people, when thinking about Geography as a subject, probably think of maps. Maps are the "where" of geography, the locational or spatial component. I tell my students that "where" is one of the first things that we learn as babies. We want to know where mommy is, where the light is, where the door to the room is and so on. Location is one of the first things that students learn when they undertake the study of a region. Where the communities are, where the routes of the roads and the rivers run are the sorts of things that must be known before a true understanding of an area is possible. Location is the information a geographer endeavours to find right from the beginning. This book was largely conceived with "where" as the main focus of its questions.

As any school year progresses, a geography teacher will shift the focus from the "where" of something to the "how" of something. Topics like how steel is made, when added to where the raw materials come from, bring about a better understanding of the subject being studied. I've written many "how" questions in this book.

Finally, I tell my students that the most comprehensive geographical understanding is arrived at when they can answer "why." Why is the steel plant located where it is? Why is Halifax where it is? Thus a variety of "why" questions are included too.

I've often called Geography the subject of borrowers. In any given geography class, it's the rare teacher who doesn't borrow ideas and information from subjects as diverse as Geology, Economics, History, Mathematics, Physics, Anthropology, Political Science, and Psychology, to name just a few. I too have borrowed from these disciplines for a number of my questions.

Finally, a word about the organization of this book. As I was writing the questions, I categorized them into more than forty topics – sports, bridges, rivers, literature, tourism, and so on. Publishing them in this state would have made for very choppy reading, so, on the advice of my publishing "guru," Ian Wright, I've grouped these subjects into larger, more inclusive categories. Thus, under the heading Communities, we move from a series of questions on famous and not-so-famous Canadian icons to a number of questions on Canadian universities, then to a sequence of questions on memorable events which have occurred in Canadian communities over the years.

I hope that many of these questions challenge you. Perhaps having an atlas nearby would be a help. I hope you will know the answers to at least some of the questions. And if you enjoy answering these as much as I have enjoyed researching and writing them, this book's mission will have succeeded.

Doug MacLean

Quiz Book

Landforms & Climate

Atmosphere & Weather
Landscapes
Water
Landforms
Islands
Events

1. The prevailing winds along the southern coast of British Columbia come from what direction?

2. What part of Canada has the highest number of frost-free days?

3. What part of Canada receives the most sunshine annually?

4. What part of Canada receives the most precipitation?

5. What Canadian location received the most precipitation in a single year?

6. Where is the driest area in Canada?

7. Which of the following Canadian provinces does *not* have semi-arid areas?
    A. Alberta
    B. B.C.
    C. Manitoba
    D. Saskatchewan

8. What is a desert region?

9. Does Canada have any desert regions?

10. What was the highest temperature ever recorded in Canada?

11. What was the lowest temperature ever recorded in Canada?

12. And in the "hang on to your hats" category, where was the highest one-hour windspeed ever recorded in Canada?

13. Recent North American weather conditions have frequently deviated from expected norms. What is the name of the weather phenomenon most often cited as the cause of these unusual conditions?

14. A dry, warm, strong wind that blows down the eastern slopes of the Rocky Mountains in Alberta is called what?

15. Inuktitut, the language of the Inuit, includes various words such as Aniu, Kinintaq, and Masak to describe a common Arctic element. Name this common element.

16. What is the aurora borealis and what is its alternate name?

17. Where do most thunderstorms occur in Canada?

18. What part of Canada enjoys the hottest summers?

19. What part of Canada has the coldest average temperatures?

20. What part of Canada receives the least snowfall?

21. Which Canadian city of more than 1 million inhabitants is the snowiest large city in the world?

22. Which Western province receives most of its rainfall in the winter months?

23. What is Canada's largest landform region?

24. By what other names is it known?

25. What is the dominant category of mineral found in the Canadian Shield?

26. How many provinces are touched by the Canadian Shield?

27. The Great Plains of the United States are called what in Canada?

28. What and where is the Hudson Plains?

29. What is the major class of rock found in the Prairies?

30. Name the 750,000 km^2 area stretching across southern Alberta, Saskatchewan, and Manitoba and into the U.S. that is pitted with millions of depressions which fill with water in the spring from melting snow and rain.

31. What is the theory of Continental Drift?

32. What part of Canada is world-renowned for the quantity, variety, and quality of its dinosaur fossils?

33. Most Quebecers live in a physiographic or landform region which lies between the Canadian Shield and the Appalachian Highlands. Name this physiographic region.

34. What is a physiographic region?

35. What is the oldest physiographic region in Canada?

36. What rock type forms the bedrock of the Great Lakes-St. Lawrence Lowlands (the region where most Canadians live)?

37. Is there a basement rock below this bedrock?

38. What two provinces and one territory are found in the Cordilleran physiographic zone?

39. The Appalachians can be found in which five Canadian provinces?

40. The term *Cordillera* is used in connection with which physical feature?
    A. Chains of mountains
    B. Groups of islands
    C. Spanish fishing trawlers
    D. Shorelines and beaches

41. Which is the older of the two southern mountain regions: the Cordillera or the Appalachians?

42. In which province would you find the Smallwood Reservoir?

43. Lake Nipigon and Lake Nipissing are often confused. Which is farther west?

44. Which of the Great Lakes is the most shallow?

45. Which of the Great Lakes has the smallest surface area?

46. Which is the largest of North America's Great Lakes?

47. Which Great Lake is farthest above sea level?
    A. Huron
    B. Superior
    C. Erie
    D. Ontario

48. Which of the Great Lakes is situated entirely in the U.S.?

49. The *kayak* was first used for hunting and fishing by ancestors of people who lived in lands bordering *what* ocean?

50. What is the largest lake in Manitoba?

51. What is the deepest lake in Canada?

52. Which of the following coastal waters has the greatest tidal range?
   A. The Bay of Fundy
   B. The Gulf of St. Lawrence
   C. Hudson Bay
   D. Northumberland Strait

53. Members of the Group of Seven frequently painted the northern shore of what Great Lake?

54. What water "highways" did the first European trans-Canadian explorers use to traverse the country?

55. What distinctly Canadian watercraft did they utilize?

56. Voyageurs were important workers in an early Canadian industry. Name the industry.

57. True *or* False – Canada has the greatest amount of fresh water in the world.

58. What percent of the world's fresh water is found in Canada?
   A. 4%
   B. 9%
   C. 13%
   D. 15%

59. Canadians use *what* percent of their annual renewable water supply?
   A. 1.5%
   B. 5.5%
   C. 9.5%
   D. 13.5%

60. Name the sector that uses the most water?
   A. Thermal power generation
   B. Manufacturing
   C. Agriculture
   D. Households

61. What province utilizes the most water?

62. Thermal power generation plants use water to cool condensers and to drive generators. The thermal power generation industry utilizes what percent of the total annual water used in Canada?
   A. 43%
   B. 53%
   C. 63%
   D. 73%

63. Do all of the Grand Banks fall within Canada's 370 km territorial limits (measured from Newfoundland)?

64. Melville Peninsula and Baffin Island form the shoreline of a large body of water known as a basin. What is the name of this body of water?

65. How many lakes are in Canada?

66. The Gulf Stream ocean current flows along what Canadian provinces?

67. What percent of the Arctic Ocean is free of ice during peak summer months?

68. If you encounter *Muskeg* in the arctic summer, you should:
    A. Play dead
    B. Put on your sunglasses
    C. Take its picture
    D. Walk carefully around it

69. What is *Muskeg?*
    A. A northern swamp
    B. A barrel-maker's tool
    C. A species of fish
    D. A shaggy, hollow-horned ruminant

70. Which province has a salt-water coastline – Manitoba or Alberta?

71. What is a *Polynya?*
    A. A glacial deposit
    B. An Arctic flower
    C. A patch of water in the Arctic Ocean which is open year round
    D. A mountain stream

72. Spear, Race, and Dorset are associated with what kind of land features?

73. What major American city is located near Niagara Falls, Ontario?

74. Lake Ontario is to Toronto as *what* is to Cleveland?

75. What do you call the seasonal ice formed by the convergence of several ice floes?

76. What do you call a large mass of ice that moves slowly over land?

77. What is a *crevasse?*

78. What region of Canada has the most glaciers?

79. Where is the popular Columbia Icefield found?

80. Columbia Icefield meltwater flows into three different oceans. Name them.

81. What is the name of the glacier tourists can traverse in specially constructed buses?

82. What is the name of mushroom-shaped rock formations created by wind erosion in dry regions, and what Canadian region is noted for these formations?

83. What is a peninsula?

84. On what island is the Avalon Peninsula?

85. On what island are Wollaston Peninsula and Collinson Peninsula?

86. On what island is Esowista Peninsula?

87. What is a *nunatak*?

88. Where in Canada are *nunataks* located?

89. Horn, Arete, Col, and Pass are words associated with what type of landform?

90. Where is Chilkoot Pass?

91. How many mountain passes between Alberta and B.C. are occupied by highways or railroads?

92. What is the highest mountain in all the provinces?

93. The Selkirk, Cariboo, Purcell, and Monashee Mountains are parts of what larger mountain system?

94. Cape Dorset is a settlement on what Canadian island?

95. What Western Canadian provincial capital is located on an island?

96. Name the largest island in the Great Lakes.

97. Name the three Eastern Canadian capitals located on islands.

98. Name the Canadian metropolis located on and covering most of an island.

99. Name the two islands in the Gulf of St. Lawrence that represent France's last colonial holdings in North America.

100. The Belcher Islands are located in what body of water?

101. The fourth largest island in the world is Canada's largest island. What island is it?

102. Name Canada's most northern island.

103. Name the large, sparsely-inhabited Quebec island located in the Gulf of St. Lawrence.

104. What is the largest of B.C.'s Queen Charlotte Islands?

105. On what island is Pacific Rim National Park?

106. Which is larger – P.E.I. or Vancouver Island?

107. The late American president, Franklin Delano Roosevelt, had a summer home on what Bay of Fundy island?

108. What large Arctic island was named after a British monarch?

109. An Arctic archipelago was named after another British monarch. Name this island chain.

110. How many Canadian Arctic islands are named after royalty, past or present?

111. What Arctic peninsula was named after a wealthy gin maker and where is this island located?

112. Iqaluit is the capital of the new Arctic territory of Nunavut. On what island is Iqaluit?

113. Are there really 1,000 islands in the Thousand Islands chain?

114. What is the name of the low-lying island group located in the Gulf of St. Lawrence? (Hint: it is part of Quebec and if you were to travel there by automobile you would have to take a ferry from Souris, P.E.I.)

115. What is the island off Nova Scotia known as the "Graveyard of the Atlantic?"

116. What animals accidentally found a home there and now thrive in its harsh climate?

117. What recent find near this island might fire up Nova Scotia's economy?

118. Prince Albert Sound and Hadley Bay are part of what Arctic island?

119. Which Canadian island has the largest human population?

120. What island off the east coast of Vancouver Island is considered a haven for artists and others escaping "the rat race"?

121. On average, how many earthquakes occur in Canada each year?

122. How many of these are strong enough to be felt by people?

123. What two parts of Canada receive the most earthquakes?

124. A famous 1903 landslide "buried a town." Name the town.

125. Approximately how many tornadoes touch down in Canada each year?

126. Where do most Canadian tornadoes occur?

127. Name the two cities hit by the Prairies' two most deadly tornadoes?

128. A 1946 tornado killed 17 people in what area of southwestern Ontario?

129. A 1985 tornado killed 12 people and caused over $100 million in property damage. What Ontario city did this tornado devastate?

130. What disaster occurred in January, 1998, and caused the most damage in Canadian history, as determined by insurance claims?

131. Hurricane Hazel struck what city in 1954?

132. What caused most of Hurricane Hazel's damage?

133. Was Hazel the first hurricane to hit Canada?

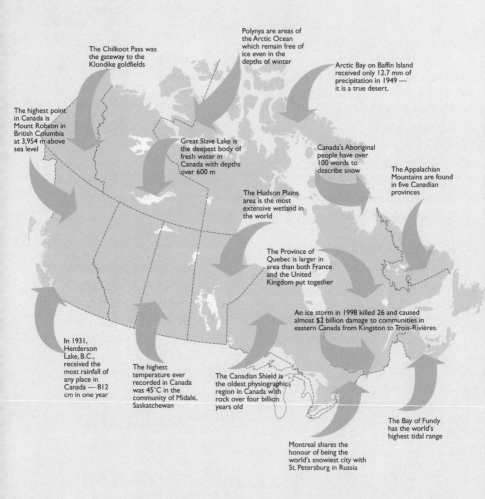

Landforms & Climate

Answers

The Chilkoot Pass was the gateway to the Klondike goldfields

Polynya are areas of the Arctic Ocean which remain free of ice even in the depths of winter

Arctic Bay on Baffin Island received only 12.7 mm of precipitation in 1949 — it is a true desert.

The highest point in Canada is Mount Robson in British Columbia at 3,954 m above sea level

Great Slave Lake is the deepest body of fresh water in Canada with depths over 600 m

Canada's Aboriginal people have over 100 words to describe snow

The Appalachian Mountains are found in five Canadian provinces

The Hudson Plains area is the most extensive wetland in the world

The Province of Quebec is larger in area than both France and the United Kingdom put together

An ice storm in 1998 killed 26 and caused almost $2 billion damage to communities in eastern Canada from Kingston to Trois-Rivières.

In 1931, Henderson Lake, B.C., received the most rainfall of any place in Canada — 812 cm in one year

The highest temperature ever recorded in Canada was 45˚C in the community of Midale, Saskatchewan

The Canadian Shield is the oldest physiographic region in Canada with rock over four billion years old

The Bay of Fundy has the world's highest tidal range

Montreal shares the honour of being the world's snowiest city with St. Petersburg in Russia

1. *West.*

2. *The number of frost-free days is the average number of consecutive days between the last killing frost of the spring and the first of the autumn. This average allows us to calculate when plants can be set out and safely harvested before the next frost. All of Canada is envious of southwestern B.C., with an average annual total of over 180 frost-free days. In Windsor, Ontario, and surrounding environs, the average is between 140 and 180 days.*

3. *Southern Saskatchewan, and southeastern Alberta each receive approximately 2,400 hours of effective sunshine annually. By contrast, Halifax and Vancouver receive about 1,800 hours, and Toronto about 2,000.*

4. *The northern coast of B.C. and the mountains of Vancouver Island. Prince Rupert, for example receives 2,400 mm of precipitation a year, more than three times that of Toronto's 800 mm yearly average.*

5. *Henderson Lake, B.C., which received 812 cm in 1931, a respectable total in any Asian monsoon area.*

6. *The Far North – precipitation, like temperature, gradually decreases from the 800 – 1,000 mm range of southern Ontario and the 500 mm of southern Manitoba to around 300 mm along the southern border of the Northwest Territories. This decrease continues until annual totals of less than 100 mm are reached on northern parts of Banks and Ellesmere Islands. One could say that precipitation decreases as latitude increases. Snowfall in the Far North is about a third of the amount received in eastern areas of Atlantic Canada.*

7. *C. Manitoba*

8. *Any region that averages less than 250 mm of annual precipitation is generally considered desert.*

9. *Arctic Bay, on northern Baffin Island, Northwest Territories, received only 12.7 mm in 1949, giving Canada world-class rank with any desert country in the world. Canadians are inclined to forget that the Arctic territories comprise one of the largest deserts in the world. There is also another desert region in the rainshadow of the Coast Mountains near Cache Creek, B.C.*

10. *45°C at Midale in southern Saskatchewan.*

11. *–63°C at Snag in the Yukon.*

12. *201 km/h at Cape Hopes Advance on Ungava Bay, Quebec.*

13. *El Niño.*

14. *A Chinook.*

15. *Snow.*

16. *Aurora Borealis, or Northern Lights, occur when charged subatomic particles from solar flares enter the Earth's upper atmosphere – to interact with oxygen and nitrogen. The resultant shimmering colours in the night sky are an unforgettable sight.*

17. *The southern Prairie provinces, the southern interior of B.C., and southern Ontario each average more than 30 thunderstorm days per year.*

18. *The Windsor-Chatham area of southern Ontario which has an average July temperature of 22°C or higher.*

19. *The northern half of Ellesmere Island, Axel Heiberg Island, and an area south of the Boothia Peninsula on the mainland of the Northwest Territories. Each of these Arctic regions boasts average January temperatures of –35°C or lower.*

20. *This honour goes to two areas: the Arctic islands because cooler air holds very little water vapour and low temperatures cause little evaporation and sublimation; and southwestern B.C. because most of this area's precipitation falls in the form of rain.*

21. *Montreal shares this dubious honour with St. Petersburg, Russia. Each city averages more than 250 cm a year.*

22. *B.C.*

23. *The Canadian Shield.*

24. *"The Shield," as it is commonly known, is also called the Precambrian Shield, in reference to the geologic era in which its rock was formed. It is also referred to as the Laurentian Shield because its southern edge is located in the Laurentian Mountains.*

25. *Metallic minerals such as gold and silver.*

26. *Six – Alberta, Saskatchewan, Manitoba, Ontario, Quebec, and Newfoundland.*

27. *The Prairies.*

28. *Located in the southern Arctic, the Hudson Plains area is*

the largest extensive wetland in the world, with shallow, open waters and peatbogs.

29. *Sedimentary rock.*

30. *The Prairie Pothole wetlands – these small bodies of water vary considerably in size and depth. Some of the larger bodies become lakes or ponds but the smaller depressions form innumerable temporary sloughs or potholes, many of them drying up in only a few weeks.*

31. *200 million years ago the continents were all grouped together as one huge landmass which we call Pangaea. The continental drift theory holds that Pangaea slowly broke up and separated into what we now know as the continents. These are still moving across the earth's surface on a substratum of magma. "Plate tectonics" is now more commonly used to refer to this movement.*

32. *Without a doubt, the Badlands of Alberta, home to both the Royal Tyrrell Museum of Paleontology at Drumheller – one of the most interesting and informative museums on dinosaurs in the world – and Dinosaur Provincial Park, which was designated a World Heritage Site by UNESCO.*

33. *The St. Lawrence Lowlands of the Great Lakes-St. Lawrence Lowlands.*

34. *It is a region of the earth's surface determined by the age and type of its rock, the forces that shaped the rock, and the surface features. Canada is subdivided into eight regions, each with its own distinctive characteristics.*

35. *Based on the age of rock, it's certainly the Canadian Shield. The age of Shield rock varies by billions of years, but an area just east of Great Bear Lake is now thought to be 4 billion years old, making it the oldest discovered rock on earth.*

36. *The major rock category is sedimentary, and the most common rock type within this category is limestone often with sandstone underlying it.*

37. *Igneous rock, seen occasionally in outcrops and the same type seen in the Canadian Shield.*

38. *B.C., Alberta, and Yukon Territory.*

39. *Quebec, New Brunswick, P.E.I., Nova Scotia, and Newfoundland.*

40. *A. Chains of mountains*

41. *The Cordillera are estimated to be 60 million years old while the Appalachians are usually said to be 400 million years old.*

42. *This reservoir is part of the Churchill River power complex in western Labrador, Newfoundland.*

43. *Lake Nipigon.*

44. *Lake Erie with a maximum depth of 65 m and an average depth of less than 20 metres.*

45. *Lake Ontario with a surface area of 18,960 km².*

46. *Lake Superior with a total surface area of 82,410 km².*

47. *B. Superior Lake Superior is 183 m above sea level*

48. *Lake Michigan.*

49. *Arctic Ocean.*

50. *Lake Winnipeg.*

51. *Great Slave Lake in the Northwest Territories. It runs to a depth of 614 m.*

52. *A. The Bay of Fundy*

53. *Lake Superior.*

54. *Rivers and lakes.*

55. *Canoe.*

56. *The fur trade.*

57. *True – Canada's fresh water bodies cover 754,000 km².*

58. *B. 9%*

59. *A. 1.5%*

60. *A. Thermal power generation*

61. *Ontario, which accounts for just over 63% of Canada's total annual water withdrawals.*

62. *C. 63%*

63. *No, while most of the Grand Banks lies within Canada's jurisdiction, approximately 30% is found in international waters.*

64. *Foxe Basin.*

65. *It is estimated that Canada has 2 million lakes.*

66. *The Gulf Stream flows off the coast of Nova Scotia, the southeastern coast of Newfoundland and on towards western Europe.*

67. *Between eight and twelve percent.*

68. D. *Walk carefully around it*

69. A. *A northern swamp*

70. *Manitoba.*

71. C. *A patch of water in the Arctic Ocean which is open year round*

72. *Capes.*

73. *Buffalo, New York.*

74. *Lake Erie.*

75. *Pack ice or ice pack.*

76. *A glacier.*

77. *A crevasse is a long open crack found at the surface of a glacier caused by stretching and ice movement.*

78. *Northern Canada by far. Ellesmere Island, Devon Island, and Baffin Island are spotted with glaciers.*

79. *In Jasper National Park near the border of Banff National Park.*

80. *The Pacific, Arctic, and Atlantic. The Columbia Icefield sits* on the Great Divide (B.C./Alberta border), which separates the flow of meltwater from the Columbia Glacier into the west through B.C. and into the Pacific, and into the east towards the Arctic and Atlantic Oceans. Part of the Alberta portion flows eastward into the Saskatchewan River, then into Hudson Bay and eventually into the Atlantic. The rest flows into the Athabaska River, Lake Athabaska, and then into Great Slave Lake, the Mackenzie River and then to the Arctic Ocean.

81. *Athabaska Glacier, located in the Columbia Icefield, just inside the entrance to Jasper National Park.*

82. *Hoodoos. Alberta, especially the areas around Banff and Drumheller, is best known for its hoodoos.*

83. *A land feature surrounded by water on all sides but one.*

84. *The eastern part of Newfoundland.*

85. *Victoria Island in the Northwest Territories.*

86. *Vancouver Island near Tofino in Pacific Rim National Park.*

87. *A mountain peak completely surrounded by glaciers, or more basically, an island of rock, soil, and vegetation in a sea of ice.*

88. *Nunataks are features of high altitude or high latitude, or most commonly, a combination of both. They are found in the mountains of Northern Canada. Some of the most studied nunataks in the world are located in the St. Elias Mountains of the Yukon.*

89. *Mountains or Mountain Passes.*

90. *This pass leads through the Coast Mountains between Skagway, Alaska, and Whitehorse in the Yukon. It was once travelled by fortune seekers during the Klondike Gold Rush.*

91. *Four. North to south they are: Yellowhead, Kicking Horse, Vermillion, and Crowsnest.*

92. *Mt. Robson in B.C. reaches 3,954 m.*

93. *The Columbia Mountains of the Western Cordillera Mountain System.*

94. *Baffin Island.*

95. *Victoria, B.C.*

96. *Manitoulin Island.*

97. *Charlottetown, P.E.I., St. John's, Newfoundland, and Iqaluit, Nunavut.*

98. *Montreal, in the St. Lawrence River.*

99. *St. Pierre and Miquelon, outside the Caribbean area.*

100. *Hudson Bay.*

101. *Baffin Island. At 507,451 km^2, it is more than twice as big as the next largest, Victoria Island, and more than 90 times the size of P.E.I.*

102. *Ellesmere Island.*

103. *Anticosti Island.*

104. *Graham Island.*

105. *Vancouver Island.*

106. *Vancouver Island.*

107. *Campobello Island.*

108. *Victoria Island.*

109. *Queen Elizabeth Islands.*

110. *Six: Queen Elizabeth Islands, Prince Patrick Island, Victoria Island, Prince of Wales Island, King William Island, and Prince Charles Island.*

111. *Boothia Peninsula, located in Nunavut between Franklin Strait and the Gulf of Boothia, was named after Briton Felix Booth.*

112. *Baffin Island.*

113. *Officially, there are 1,149 islands – 665 are Canadian and 484 are American. Almost half are so small (islets) that they are not even named.*

114. *Magdalen Islands or Îles de la Madeleine.*

115. *Sable Island which has claimed more than 300 ships over the years.*

116. *Ponies which swam ashore from some of the many ships wrecked along the shore. Sable Island provides a harsh but predator-free environment for these unique creatures.*

117. *Oil and natural gas.*

118. *Victoria Island.*

119. *Montreal, with a population of almost 2 million people.*

120. *Saltspring.*

121. *Almost 1,500 earthquakes occur in Canada annually.*

122. *An average of 50.*

123. *These two regions are easy to identify on a map of past earthquakes: coastal B.C. and the lower St. Lawrence River Valley in Quebec. Landslides in the clay sub-soil of Quebec have been common and costly consequences of earthquakes. The 1971 St. Jean Vianney slide, for instance, left 31 people dead. Layers of soil covering the clay slid downhill after a small earthquake and carried dozens of houses and residents into the St. Jean Vianney River. Over 800 scars of earlier landslides can be seen on air photos of the region.*

124. *The coal-mining town of Frank, Alberta, which sat at the foot of Turtle Mountain. At 4:00 a.m. on April 29, a small earthquake dislodged a huge, 60 million tonne chunk from Turtle Mountain which slid down over the town. More than 100 people were killed.*

125. *Usually less than a dozen, but because many occur in isolated, even uninhabited regions, the real number is unknown.*

126. *Most tornadoes occur in the Prairies and southwestern Ontario. They are not, however, restricted to these regions as they* do occasionally occur in the Shield region. People working on Sudbury, Ontario's, Inco super-stack in 1970, for instance, recall looking down at a tornado that caused $10 million in damage and killed 6 people.

127. The Regina Tornado of 1912 left 28 people dead and many injured. The 1987 Edmonton Tornado killed 27 people and caused $300 million damage.

128. It touched down between Windsor and Tecumseh.

129. Barrie.

130. The Ice Storm affected much of eastern North America and covered the region stretching eastward from Kingston to Trois Rivières with over 100 mm of ice. According to the Insurance Bureau of Canada, ice storm related claims totaled almost $2 billion and caused 22 deaths in Quebec and 4 in Ontario.

131. The hurricane struck Toronto; 81 people were killed. The disaster was unusual because hurricanes are most typically a subtropical coastal phenomenon associated with high winds, surging tides, and heavy rain. Toronto is certainly not subtropical, is located far from the ocean, and is usually protected from Atlantic hurricanes by the Appalachian Mountains which serve as a barrier.

132. Eighteen centimetres of rain in twenty-four hours, preceded by several weeks of higher than usual rainfall. The early rainfall saturated the soil. When the hurricane rain fell, the water could not be absorbed, and ran off the soil filling ditches and streams, turning these into raging torrents. Twenty-four of the twenty-eight bridges in the affected area were destroyed.

133. Certainly not! Maritimers know all about the "October Gales." The most destructive of these "tail-ends" of hurricanes hit Cape Breton on August 25 , 1873, and left almost 1,000 people dead, some 1,200 ships sunk or smashed, and hundreds of homes and bridges destroyed. Sadly, meteorologists in Toronto knew a day early that the hurricane would strike the Maritimes, but downed telegraph lines prevented the transmission of this information.

Resources

Agriculture
Parks
Animals
Forests
Tourism
Food
Mining
Legend
Events

1. Which Canadian province produces twice as much wheat as all nine others combined?
   A. Alberta
   B. Manitoba
   C. Ontario
   D. Saskatchewan

2. Which of the following Western crops is *not* used to produce an edible oil?
   A. Canola
   B. Flax
   C. Millet
   D. Sunflowers

3. Ontario's Niagara Peninsula is famous for its fruits and grapes. What other large, Western agricultural region in Canada is famous for its fruit?

4. Name the low cholesterol oil seed crop invented in Canada that is in constant demand around the world.

5. Which Canadian province has long been referred to as the "Million Acre Farm"?

6. What Manitoba research centre was founded in 1915 as an experimental tree nursery and steer feeding farm?

7. Maize, as it is known throughout the world, is better known to Canadians as *what*?

8. In 1995, the Canadian government ended a rail subsidy that allowed for reduced freight rates on eastbound shipments of Western grain and flour. This subsidy was part of an agreement named after a Rocky Mountain pass. Name the agreement.

9. Which province has the greatest dairy production in Canada?
   A. P.E.I.
   B. Ontario
   C. Alberta
   D. Quebec

10. Which of the following farm products generates the most value in Canada?
   A. Cattle
   B. Dairy products
   C. Wheat
   D. Pigs

11. Which province receives the greatest financial return from its agricultural produce?
A. Quebec
B. Saskatchewan
C. Alberta
D. Ontario

12. Where are Canada's largest farms?

13. Which province has the most farms?

14. What part of Canada has the longest growing season?

15. Does Canada import or export more food, in dollars?

16. Where was that most Canadian of apples, the McIntosh, discovered?

17. What are "fiddleheads"?

18. Where are fiddleheads found?

19. Which province is the leading producer of maple syrup?

20. How many national parks are in Canada?

21. Jasper National Park is to Alberta as Kluane National Park is to *where?*

22. Which of the following four national parks receives the most visitors per year?
A. Banff, Alberta
B. Wapusk, Manitoba
C. Point Pelee, Ontario
D. Kootenay, B.C.

23. Grasslands National Park is located in which Canadian province?

24. Canada's first national park was founded in November of 1885. Name this Western park.

25. The above park was founded 13 years after the first U.S. national park. Name this park.

26. Dinosaur Provincial Park contains Upper Cretaceous dinosaur fossils and was declared a World Heritage Site by the United Nations in 1979. In which Canadian province is it located?

27. Banff National Park is to Alberta as Prince Albert National Park is to...?

28. Kluane National Park is to the Yukon as Nahanni National Park is to...?

29. Cape Breton Highlands National Park is to Nova Scotia as *what* is to Newfoundland?

30. What spectacular land-forms are associated with the above Newfoundland park?

31. All the statements in the following are true but one. Locate the error.

Alberta's Banff National Park was established in 1885 to pro-tect public access to the province's famous mineral hot spring. Since 1885, Canada's first national park has grown to more than 250 times its original size and includes some of the finest alpine scenery in the world including Mt. Robson, Canada's tallest mountain. The area is also noted for its wide variety of wildlife including Rocky Mountain goats, bighorn sheep, elk, moose, deer, and both black and grizzly bears. Banff has, for many years, attracted more than a million tourists annually. Its accommodations range from luxurious hotels at Banff and Lake Louise to campgrounds sprinkled throughout the park. Along the scenic Icefields Highway connecting Banff and

Jasper National Parks is the Columbia Icefield, the largest of its kind in North America.

32. In what province would you find Riding Mountain National Park?

33. Terra Nova National Park is located in which province?

34. If you were visiting the fortress at the Louisbourg National Historic Site, what province would you be in?

35. Aboriginal groups blocked off areas of land in what two provinces during the summer of 1995?

36. In what province would you find Batoche National Historic Site?

37. What important historical event occurred at Batoche National Historic Site?

38. In what national park is Radium Hot Springs?

39. The following describes what National Park: "Scenic Cabot Trail characterized by a rugged shoreline with plunging cliffs."

40. On which Arctic island is Auyuittuq National Park? What does the name mean?

41. In which province does one find Kejimkujik National Park?

42. In which province is Kouchibouguac National Park? What does its name mean, and from what language is it derived?

43. Which of Canada's national parks is found farthest north?

44. Which national park is found farthest east?

45. Which national park is found farthest west?

46. Which national park is farthest south?

47. Where is Fathom Five National Park? When was it established and why?

48. Where is Pukaskwa National Park?

49. Where is the "Sleeping Giant"?

50. What attracts tourists to Wapusk National Park?

51. In 1992, the Canadian government called a stoppage – or a moratorium – on the fishing of a specific species of fish off Canada's East Coast. Name the species of fish.

52. Canada (and other sea-coast countries) is allowed to protect and manage fish stocks over how many kilometres from its coastal shore?
A. 3 km
B. 37 km
C. 500 km
D. 370 km

53. Which of the following countries catches more fish, by weight, than Canada?
A. Chile
B. France
C. United Kingdom
D. Brazil

RESOURCES

54. How much of Canada's seafood catch is exported?
   A. 20%
   B. 40%
   C. 60%
   D. 80%

55. To which nation does Canada send most of its seafood catch?
   A. Spain
   B. United States
   C. United Kingdom
   D. Japan

56. What percent of the primary sector's contribution to Canada's gross domestic product comes from the fishing industry?
   A. 2%
   B. 28%
   C. 16%
   D. 12%

57. How many commercial fishermen are there in Canada?
   A. 23,000
   B. 33,000
   C. 43,000
   D. 53,000

58. How many commercial fishermen were there before the Atlantic fishing moratorium was enacted in 1992?
   A. 21,000
   B. 31,000
   C. 41,000
   D. 51,000

59. The original 1992 moratorium on East Coast fishing was extended indefinitely in 1994 because of alarmingly low numbers of remaining cod. What percent of late 1980's totals were left in 1994?

60. What caused this deplorable situation?

61. Salmon fishing has been restricted or banned in various parts of Canada due to a reduction in fish stocks and a concern for the future of the salmon fishery. Which province has suffered most from these limitations?

62. How many Canadian animal and fish populations have become extinct since the advent of European settlement?
   A. 5
   B. 20
   C. 100
   D. 200

63. How many wildlife species are threatened with extinction in Canada?
- A. 17
- B. 117
- C. 170
- D. 1,070

64. Which of the following Canadian animal types is *not* extinct?
- A. Dawson Caribou
- B. Sea mink
- C. Passenger pigeon
- D. Whooping crane

65. What cold water fish is commonly known in various regions of Canada by such names as Doré, Mooneye, Pickerel and Pike-perch?

66. Which of these fish is *not* a species that was introduced to the Great Lakes?
- A. Brown Trout
- B. Carp
- C. Coho Salmon
- D. Lake Trout

67. What percent of Canadians list recreational fishing among their pastimes?
- A. 7%
- B. 18%
- C. 30%
- D. 42%

68. What percent of Canadians list hunting among their pastimes?
- A. 5%
- B. 15%
- C. 20%
- D. 33%

69. What percent of Canadians list the less-consumptive wildlife related activities of wildlife photography and bird-watching among their pastimes?
- A. 6%
- B. 13%
- C. 25%
- D. 30%

70. Canadian hunters blast some 2 millions tonnes of shot into the environment each year. In addition, an estimated 500 tonnes worth of sinkers and jigs are lost in Canadian waters every year. Why is this a problem?

71. The loon, that quintessential symbol of Canada, is capable of flying 100 km/h. What is its chief flying problem?

72. What is the loon's real forté?

73. What magnificent bird has disappeared from large areas of Canada since European migration?

74. Over 1,600 of these predatory birds were raised by the Canadian Wildlife Service at a facility in Wainwright, Alberta, and released into the wild. Name this bird.

75. Huge flocks of these birds wintered in Argentina, and returned in spring to nest in the tundra of Canada's Northwest Territories. After nesting, massive flocks flew east to Labrador and Newfoundland where they fattened up on crowberries before the gruelling flight over the Atlantic to the coasts of South America. At one time these birds were hunted for sport and sale in the market. Today they are nearly extinct and fewer than 100 may survive. Name this bird.

76. A reduction in hunting and an increase in unharvested grain crops means these Arctic birds are increasing in numbers and damaging northern habitat. Name this bird?

77. What songbird is important in controlling insect pests which destroy valuable forest resources?

78. What duck's nest is harvested – not for soup – but for feathers?

79. Every spring, visitors to this tiny national park come to see the birds. Name this park.

80. Where is North America's oldest waterfowl refuge?

81. Ten thousand Canadians complete what census every Christmas?

82. What animal is Canada's largest rodent and was almost extinct in the early twentieth century?

83. What Canadian member of the cat family has a population that goes through ten-year "boom-and-bust" cycles?

84. Which one of the following Prairie species is increasing in population?
- A. Black-footed ferret
- B. Wood buffalo
- C. Burrowing owl
- D. Sprague's Pipit

85. What small, agile mammal, about the size of a housecat, was common on Canada's southern Prairies in the nineteenth century, but was designated an extirpated species in 1978?

86. The Inuit name for this animal is omingmak, "the animal with skin like a beard." What do English-speaking Canadians call it?

87. Where is the Dempster Highway and what animal-related problem is associated with it?

88. More than 160,000 caribou make up the Porcupine Caribou herd. Where are these animals to be found?

89. This woodchuck is being "chucked" out of its home – where?

90. Why does clear-cutting mean fewer bears?

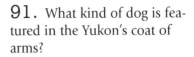

91. What kind of dog is featured in the Yukon's coat of arms?

92. Which tiny flying insect performs one of the world's great migrations?

93. What tiny mollusk has proliferated throughout the waters of the Great Lakes?

94. What colourful wetland plant is considered a noxious invader and costs the country thousands of dollars in clean-up each year?

95. A tiny fungus creates a disease that almost eradicated one of North America's favourite trees. Name the disease.

96. Salicin is an ingredient in Aspirin and is derived from what common wild tree?

97. What type of vegetation is found in a parkland zone?

98. This zone is more commonly known by what name in the rest of the world?

99. In which province would you find the largest area of broad leafed trees or Carolinian vegetation?

RESOURCES

100. Silviculture refers to what economic activity?

101. Which of the four major tree harvesting techniques is most used in the Canadian forestry industry?
A. Seed tree cutting and re-planting
B. Shelterwood cutting
C. Single tree selective cutting
D. Clear-cutting

102. What percent of harvested forest land was clear-cut?
A. 55% — 65%
B. 65% — 75%
C. 75% — 85%
D. 85 — 95%

103. What term is used to describe a climax forest consisting of old and usually large trees?

104. Which province has the most forest land?

105. Which province has the largest volume of forest wood?

106. Why is Canada's temperate rainforest ecozone so important?

107. The following countries lead the world in forest product exports. Rearrange the list from largest exporter to smallest exporter.
A. Sweden
B. Finland
C. Canada
D. United States

108. What country is the main recipient of Canada's forest product exports?

109. What are Canada's three leading forest products?

110. Approximately how many Canadians are directly employed in the forestry industry?
A. 100,000
B. 200,000
C. 300,000
D. 400,000

111. Approximately how many Canadian communities depend wholly or partially on the forestry industry?

A. 100
B. 400
C. 900
D. 2,000

112. Divide the following words into the two common tree categories:
coniferous, evergreen, hardwood, softwood, deciduous, leaf bearing, cone bearing

113. What small area of West Coast forest has become a battleground between loggers and conservationists?

114. Which of the following Canadian forest types is not considered economically productive?
   A. Boreal forest and grassland
   B. Boreal forest
   C. Boreal forest and barren ground
   D. Boreal – predominantly forest

115. What Canadian group first chose the maple leaf as its official symbol?

116. In what year did the maple leaf become the main symbol on Canada's flag?

117. What foreign country is the most popular destination for Canadian travelers?

118. After New York City, what is the biggest tourist attraction in North America?

119. Like so many other nations, tourism plays a major part in Canada's economy. Where does Canada rank in spending by tourists?

120. Which Atlantic province receives the most tourist visits per year from non-residents?

121. How much money is spent by Canadian and foreign tourists in Canada each year?
   A. 5 billion
   B. 15 billion
   C. 20 billion
   D. 25 billion

122. Who spends more money each year: visitors to Canada or Canadians visiting other countries?

123. From what country do most visitors to Canada originate?

124. What specific foreign destination region is the beneficiary of most Canadian tourist dollars?

125. Which three provinces are most visited by American tourists?

126. Americans visit us most. What countries send the second and third most visitors to Canada?

127. What city is the most popular Canadian destination for American tourists?

128. In what season do most tourists come to Canada?

129. A specific fish formed a staple for Canada's West Coast Aboriginal peoples and later became an important export product. Name this fish?

130. What is *sagamite?*

131. What is the name of the dried, powdered meat and edible berry concoction that fed the early fur traders?

132. Name the bread staple eaten by early settlers and fur trappers.

133. What is the chocolaty dessert bar named after a West Coast city?

134. From what part of Canada does a concoction of french fries and cheese curds

covered in brown gravy originate and what is it called?

135. The study of the diffusion of beer in Canada is the study of Canadian settlement patterns. Where in Canada did European settlers brew the first beer?

136. Who built the first commercial brewery and where was it built?

137. Where is the oldest continuously-operating brewery in North America located?

138. When was the second largest commercial brewery in Canada founded and where was it located?

139. Most Canadians know that our two largest breweries are Molson's and Labatt Breweries. When did Labatt's get started and where?

140. When was the first commercial brewery founded in the Prairie provinces and where was it located?

141. Where and when was B.C.'s first commercial brewery founded?

142. Which region of Canada has the highest per capita beer consumption?

143. Which Canadian region has the lowest per capita beer consumption?

144. Which Canadian distillery has the largest Canadian output and where is it located?

145. What is the name of the large distillery in Walkerville, Ontario, and when was it founded?

146. What town in Ontario was once called the "Uranium Capital of the World"?

147. Roberts Bank, a port facility near Vancouver, handles more of *what* mineral than any other port in Canada?

148. A certain mineral and Quebec town where it is mined share the same name. What is it?

149. Where was Canada's first commercial oil well located?

150. Before World War II, most Canadian oil was produced in southern Ontario. After 1947, the discovery of oil in a certain Alberta town lead to a major westward shift in oil production. Name this town and its famous well.

151. Canadian coal-mining towns have experienced many deadly disasters over the years. No town is more noted for its cave-ins than Springhill. In what province is Springhill?

152. A display mine for tourists was created as a Canadian Centennial project in Glace Bay, Nova Scotia. What mineral is mined here?

153. What part does Lake Maracaibo in Venezuela play in the Canadian oil industry?

154. What percent of Canada's mineral production is exported?
A. 20%
B. 40%
C. 60%
D. 80%

155. Canadian mining is dependent on the economic health of its trade partners. Why?

156. Approximately how many different types of minerals are mined in Canada?

    A. 20
    B. 40
    C. 60
    D. 80

157. Ranked from highest to lowest, the provinces which generate the most value through the production of minerals are:

A. Quebec – B.C. – Alberta – Ontario
B. Ontario – Quebec – B.C. – Alberta
C. B.C. – Alberta – Quebec – Ontario
D. Alberta – Ontario – B.C. – Quebec

158. Which Canadian mineral ranks highest in mining revenues?

    A. Petroleum and natural gas
    B. Gold
    C. Copper
    D. Coal

159. What is the most valuable metallic mineral mined in Canada?

160. Canada ranks among the top four in world production of copper. Half of this metal is used to produce what type of product?

161. In 1994, a major nickel and copper deposit was discovered at a place called Voisey Bay. In what province or territory is Voisey Bay?

162. Canada is one of the world leaders in the production of aluminum, yet no aluminum ore (bauxite) is mined in Canada. Why does Canada have so many aluminum smelters and no aluminum mines?

163. Which province is home to the most aluminum smelters?

164. Arvida, located in the Chicoutimi-Jonquière area conurbation of Quebec, is one of the world's great aluminum producers. Where did the town of Arvida get its name?

165. The history of mining in Canada is a history of strikes and labour struggle. One very famous miners' strike took place in 1949, pitting miners and their families against the mining company and a pro-employer government. Where did the strike take place?

166. This strike helped build the political careers of what two famous Quebec politicians?

167. Strikes paralyzed what Western city for almost six weeks in the early summer of 1919?

168. Quebec produces much more electricity than it consumes. Where does the excess go?

169. What two Canadian cities have "streets paved with gold"?

170. The Klondike Gold Rush began soon after gold was discovered in what creek?

171. True *or* False – For two centuries, European demand for beaver skins and other pelts encouraged the exploration of North America.

172. Sealing has long been a spring-time occupation of East Coast fishermen.

Airplanes, helicopters, two-way radios, large steel ships and the use of Global Positioning Systems (GPS) have recently taken much of the danger out of the ice-bound hunt which is often done during harsh spring weather. In earlier days, sealing led to many deaths. What province has produced the most sealers and sealing vessels over the years?

173. Where does the seal hunt take place?

174. What tragedy took place during the seal hunt of 1914?

175. Canada is a world leader in total number of annual forest fires. Where did Canada's most deadly forest fire occur?

176. Two other major Northern Ontario fires occurred in the early part of the twentieth century. Where were they?

177. Which causes greater damage to forests, fire or insect infestation?

Resources
Answers

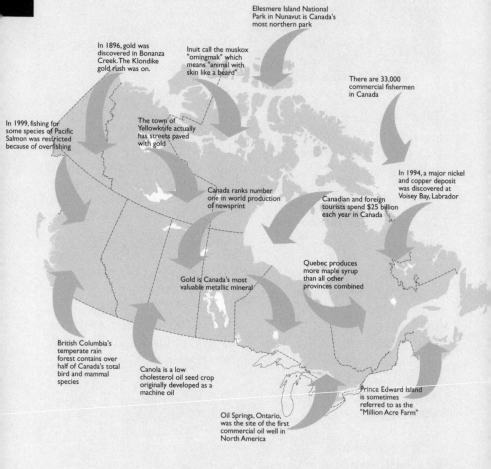

Ellesmere Island National Park in Nunavut is Canada's most northern park

In 1896, gold was discovered in Bonanza Creek. The Klondike gold rush was on.

Inuit call the muskox "omingmak" which means "animal with skin like a beard"

There are 33,000 commercial fishermen in Canada

In 1999, fishing for some species of Pacific Salmon was restricted because of overfishing

The town of Yellowknife actually has streets paved with gold

In 1994, a major nickel and copper deposit was discovered at Voisey Bay, Labrador

Canada ranks number one in world production of newsprint

Canadian and foreign tourists spend $25 billion each year in Canada

Quebec produces more maple syrup than all other provinces combined

Gold is Canada's most valuable metallic mineral

British Columbia's temperate rain forest contains over half of Canada's total bird and mammal species

Canola is a low cholesterol oil seed crop originally developed as a machine oil

Prince Edward Island is sometimes referred to as the "Million Acre Farm"

Oil Springs, Ontario, was the site of the first commercial oil well in North America

1. *D. Saskatchewan*

2. *C. Millet*

3. *The Okanagan Valley in B.C.*

4. *Canola – a genetically altered variety of rapeseed originally developed in Canada during World War II as a source of high-quality lubricating oil for marine engines. It is now used in margarine and shortening. The original rapeseed came from Asia.*

5. *P.E.I. – with a land area of 5,660 km² – is often considered one big green farm. A tour around P.E.I. in the summertime reinforces this image. However, P.E.I., like much of rural Canada, has witnessed consolidation of its smaller farms. The area being farmed doesn't look much different but there are fewer farms.*

6. *The Morden Research Centre, located in Morden on the western edge of the Red River Valley. Today, its diversified crop lands and state-of-the-art laboratories make the Morden one of the most sophisticated and best known agricultural research centres in the world.*

7. *Corn.*

8. *Crow's Nest Pass Agreement which was first passed in 1897.*

9. *D. Quebec*

10. A. Cattle
    B. Dairy products
    C. Wheat
    D. Pigs

   The most valuable products are listed in decreasing order from A to D.

11. *D. Ontario*
   Ontario leads the way, followed by Saskatchewan and Alberta. Quebec is the least successful of the four listed, producing about 70% as much as Ontario.

12. *Saskatchewan's farms are the largest, with an average farm size just under 440 ha. Alberta is next, averaging 360 ha, followed by Manitoba at 300 ha.*

13. *Ontario, with almost 70,000 farms, followed by Saskatchewan with 61,000 and Alberta with 57,000.*

14. *The area surrounding Victoria on Vancouver Island which has a growing season exceeding 260 days. A similar area around Vancouver averages over 240 days. Windsor, Ontario, has the longest growing season in Eastern Canada with 220 days.*

RESOURCES

15. *Canada exports almost 60% more food in dollars than it imports. Fruits and vegetables make up about 40% of our imports while wheat alone makes up about 40% of our exports.*

16. *It was discovered by John McIntosh, a farmer who owned land in Dundela, located in eastern Ontario. McIntosh found the apples growing wild on his property and through a process of transplantation and grafting, he managed to cultivate the "Mac."*

17. *Fiddleheads are the edible fronds of the ostrich fern that resemble the heads of violins. These delicious greens appear in early spring and are among the first fresh Canadian vegetables available.*

18. *New Brunswick is best-known for its fiddlehead production although similar edible ferns are found from the Atlantic to the Pacific, and from the Canadian Shield to the U.S. border.*

19. *Quebec produces more than all other provinces combined. Canada, as a whole, is the world's leading producer of maple syrup.*

20. *There were 38 national parks in 1999, covering over 222,700 km² or 2.2% of the entire country.*

21. *The Yukon Territory.*

22. *A. Banff, Alberta*

23. *Saskatchewan — Grasslands preserves over 40 varieties of prairie grasses, many of which are rare or endangered.*

24. *Banff National Park.*

25. *Yellowstone, which was declared a national park in 1872.*

26. *Alberta.*

27. *Saskatchewan – this park protects a vast transition zone of aspen parkland and boreal forest.*

28. *Nahanni is located in the Northwest Territories and is set deep within the Mackenzie Mountains.*

29. *Gros Morne National Park.*

30. *Fjords.*

31. *Actually, there are two errors here. The Yukon's Mt. Logan is Canada's tallest peak. Mt. Robson, on the other hand, is the tallest peak in the ten provinces. But it is not located in Banff. Rather, it is located in B.C. in Mt. Robson Provincial Park, close to Jasper. One could also argue that the above contains a third error: Alberta did* not become a Canadian province until 1905.

32. Manitoba – this park preserves an area where the boreal forest, eastern deciduous forest, and aspen parkland converge.

33. Terra Nova is located in Newfoundland and protects almost 200 km of rugged coastline dotted with fjords, coves, and tidal flats.

34. Nova Scotia.

35. Ipperwash Provincial Park on Lake Huron, Ontario, and Gustafsen Lake in the Cariboo region of B.C.

36. Saskatchewan – the site is located near Saskatoon.

37. Batoche was home to the main Métis settlement and site of the major Northwest Rebellion confrontation that saw the defeat of Louis Riel's and Gabriel Dumont's forces on May 12, 1885.

38. Radium Hot Springs is found in Kootenay National Park, B.C., at the west end of the Banff-Windermere Parkway. The Hot Springs was re-built in the early 1970's after a trucking accident spread flaming fuel into Sinclair Creek, which flows into

the Springs, and destroyed most of the complex and surroundings.

39. Cape Breton Highlands National Park.

40. The Park was established in 1976 and is found on Baffin Island facing the Davis Strait, in Nunavut. The name means "land that never melts."

41. Nova Scotia. Lakes and rivers comprise 15% of this lush 381 km² park.

42. New Brunswick, facing the Northumberland Strait. Kouchibouguac is a Mi'kmaq word meaning "river of the long tides." The outstanding features of the park include tidewater lagoons, grassy salt marshes, and bogs.

43. Ellesmere Island National Park in Nunavut.

44. Terra Nova National Park in Newfoundland.

45. Kluane National Park together with Wrangell National Park in Alaska straddle Canada's most western point, the Alaska-Yukon border, on the 141° west longitude line.

46. Point Pelee is furthest south and at 20 km² is also Canada's smallest national park.

47. Fathom Five, Canada's first national marine park, was established in 1987. It preserves a marine ecosystem of 130 km², at the tip of the Bruce Peninsula where Lake Huron meets Georgian Bay.

48. South of Marathon on the north shore of Lake Superior in Ontario. The park contains the archeologically mysterious Puka-skwa Pits, which range from slight depressions to large pits or even "lodges," some of which are even interconnected.

49. It is part of Sleeping Giant Provincial Park in Ontario. When seen from Thunder Bay, it resembles a huge reclining figure.

50. Polar bears! About 1,200 of these majestic animals gather to conceive and bear their young each year at Wapusk, southeast of Churchill, Manitoba, on Hudson Bay.

51. Cod.

52. C. 500 km

53. A. Chile
Chile's fish catch ranks third in the world with about 7 million tonnes, about four times Canada's total. France and Brazil maintain about half of Canada's total catch, and the United Kingdom is last on this list with a catch about a third the size of Canada's. A dozen years ago, both Canada's and Chile's catches were almost equal.

54. D. 80%

55. B. United States

56. A. 2%
In dollars, 2% equals $800 million.

57. B. 33,000

58. C. 41,000

59. An outstandingly low 1% of late 1980's totals was sufficient reason for the indefinite ban.

60. In the period from 1850-1950, it is estimated that the annual northern cod catch was 250,000 tonnes. In 1968, the catch peaked at 800,000 tonnes. It remained high for the next 20 years until the precipitous drop in the late 1980's. Over-fishing, mainly by gigantic off-shore trawlers, is blamed for this calamity.

61. B.C.

62. B. 20
Since about 1750, Canada has lost such species as the great auk,

passenger pigeon, Labrador duck, eelgrass limpet, Hadley Lake stickleback, Banff longnose dace, deepwater cisco, longjaw cisco, and blue walleye.

63. B. 117
There are 339 species that are considered "at risk."

64. D. Whooping crane
This large bird which makes an annual migration from Texas to Alberta and back has made something of a comeback since its numbers fell below 20 in the 1960's. Latest estimates show the number of whooping cranes to be around 200.

65. Walleye.

66. D. Lake Trout

67. B. 18%

68. A. 5%

69. B. 13%

70. Many bird species, including mallards, black ducks, northern pintails, Canada geese, snow geese, and tundra swans swallow the shot and are poisoned by the lead content. Incidentally, the soil at some clay-target shooting ranges contains enough lead to be classed as hazardous waste under Canadian guidelines.

71. Taking off – it can't take off from land and must half-run and half-fly across water to get airborne. Once airborne, however, the loon is graceful and a pleasure to behold.

72. Diving – it can stay submerged for up to three minutes and dive to a depth of 70 m.

73. The bald eagle – today its numbers are slowly increasing due to the intensive efforts of many naturalists.

74. The peregrine falcon. Thanks to the success of the Wainwright program, captive breeding is no longer necessary and the facility was recently closed.

75. The Eskimo curlew.

76. The lesser snow goose has experienced a population explosion. These birds are seriously damaging Arctic ecosystems, particularly along the Hudson and James Bay coasts.

77. The evening grosbeak's appetite for larva of the spruce budworm, a serious pest in softwood forests, makes it one of our most beneficial songbirds.

78. On the east coast of Canada, the nest of the eider duck is harvested for eiderdown which is still the best insulation material known.

79. Point Pelee National Park in southern Ontario. Some 40 species of brightly coloured warblers draw thousands of bird watchers, who spend almost $6 million a year in the area on accommodation, food, travel, and equipment.

80. Last Mountain Lake in Saskatchewan. Parliament set aside 1,000 ha at the north end of this lake to protect breeding grounds for "Wild Fowl" in 1887.

81. Canadians across the country take time each Christmas to venture into the snow for the Christmas Bird Count, an annual census created to establish the number and species of birds wintering in Canada.

82. The beaver – today it has become so numerous that it is considered a pest in some parts of Canada.

83. The Canada lynx. This small wildcat hunts snowshoe hare, whose populations expand rapidly and then crash in 10-year cycles. The Canada lynx populations fluctuate accordingly.

84. C. Burrowing owl

85. The swift fox. It lost its habitat to farmland and was also the unintended victim of trapping and poisoning campaigns aimed at other animals such as coyotes, wolves, and ground squirrels. But the swift fox has made a recent comeback. In 1997 there were an estimated 300 wild swift foxes in Canada – the result of breeding by the Canadian Wildlife Services.

86. The muskox. This arctic animal has two layers of hair: a woolly insulating layer next to the skin and a long, coarse, oily outer layer. The wool, or "qiviut," is stronger and eight times warmer than sheep's wool and finer than cashmere. The outer hair is the longest of any mammal in North America. Together these layers allow the muskox to function normally in high winds, blowing snow and temperatures of –40°C.

87. The highway runs north/south from Dawson City in the Yukon to Inuvik in the Northwest Territories. It interferes with the annual migration of the Porcupine Caribou herd.

88. The Porcupine Caribou herd migrates between north and central Yukon's Ogilvie and

Richardson Mountains, and the Arctic plains.

89. The Vancouver Island marmot (Marmota vancouverensis) lives in the mountains of Vancouver Island, B.C. and no where else. Estimates indicate that there are fewer than 100 of these woodchucks left on the planet.

90. Hundreds of thousands of salmon – coho, chinook, sockeye, pink and chum – return to the rivers and streams of B.C. to spawn and die. Clear-cutting removes shade and leads to warmer river and stream water; it allows silt and mud to fill the crevices in the gravel spawning beds. Logging roads dam up stream flows and alter run-off rates. All these factors mean fewer salmon spawn successfully. These fatty fish provide a major food source for the great bears of the West Coast. Fewer fish – fewer bears.

91. Malamute.

92. The monarch butterfly travels from sun-drenched Canadian fields to over-wintering sites in northern Mexico every year.

93. The zebra mussel, introduced into Lake Erie in 1988, is considered a dangerous menace, despite our cleaner lakes.

94. Purple loosestrife.

95. Dutch elm disease.

96. Salicin is found in the bark of the willow tree.

97. This transitional zone found in the southern Prairies holds a mix of grassland and widely-spaced trees.

98. Savannah.

99. Ontario.

100. Commercial forestry – the science of developing tree stock and the re-planting of forests.

101. D. Clear-cutting

102. D. 85 — 95%

103. Old growth forest.

104. Quebec has over 50 million ha of forest land. B.C. is in second place with about 20% less forest area.

105. B.C. has more than half the Canadian total.

106. Some 35,000 bird and mammal species – half of all in Canada – are found exclusively in B.C. The rare white black bear variation, called Kermode or

RESOURCES

Spirit Bears, Roosevelt elk, wolf, grizzly bear, marbled murrelet and marten, all depend on the rainforest in order to survive.

107.
1. Canada — 21% of world exports
2. United States — 13% of world exports
3. Finland — 10% of world exports
4. Sweden — 9% of world exports

108. The United States.

109. Canada ranks first in the world in the production of newsprint, second in the production of wood pulp, and third in the production of softwood lumber.

110. C. 300,000

111. C. 900

112. First Group: cone bearing, coniferous, evergreen, softwood
Second Group: deciduous, hardwood, leaf bearing

113. Clayoquot Sound, on Vancouver Island, is the largest area of lowland coastal temperate rainforest left in the world. Coastal temperate rainforest covers less than 0.5% of the Earth's

surface and, according to Greenpeace Canada, 57% of the temperate rainforest in B.C. has already been clear-cut.

114. C. Boreal forest and barren ground

115. Ironically, it was the French-Canadian patriotic organization, the St.-Jean-Baptiste Society, which chose the maple leaf in 1834. The leaf, however, had long been the unofficial symbol of things Canadian for centuries prior. The St.-Jean-Baptiste Society now flies the Fleur-de-Lis.

116. The flag as we now know it replaced the Red Ensign in 1965.

117. The United States.

118. Niagara Falls, which receives over 10 million visitors a year.

119. Canada ranks 9th in world tourism dollars. The United States is in first place. Countries two through eight inclusive are all European.

120. New Brunswick. Most travellers arrive by automobile and anyone travelling to P.E.I., Nova Scotia, or Newfoundland from "Upper Canada" or the United States must drive through

New Brunswick first. Nova Scotia receives 80% of the tourists New Brunswick does, P.E.I. 40% of New Brunswick's total, and Newfoundland receives 20%.

121. D. 25 billion

122. Each year, Canadians spend about $16 billion visiting other countries. Visitors to Canada, on the other hand, spend only $10 billion here. Our weak dollar is reversing this trend making Canada more attractive to outsiders and citizens alike.

123. The United States. Americans constitute 80% of all non-Canadian visitors.

124. Except for the diffused amount spent on short trips to states on the Canadian border, the answer is Florida, which claims between $2 and $3 billion annually (about 30% of our total tourist dollars spent in the United States).

125. The top three provinces in order are: Ontario, B.C., and Quebec.

126. A distant second is the United Kingdom, followed by Japan.

127. Toronto.

128. More visitors arrive in summer than all the other seasons combined.

129. Salmon.

130. Sagamite was a cornmeal porridge once eaten by Iroquois warriors when on a war march. The Iroquois grew corn, thereby maintaining a military advantage over their nomadic enemies who were forced to hunt for food when marching.

131. Pemmican – the concoction was usually made of bison meat and the most commonly available fruit, blueberries, for example.

132. Bannock, an unleavened bread of Scottish origin made of flour, water and fat.

133. The Nanaimo bar.

134. This high-cholesterol feast named poutine originated in Quebec. It has now spread throughout much of the country. One variation uses a spicy spaghetti sauce in place of the brown gravy.

135. The first European-style beer was brewed by Jesuits in Quebec City in 1646.

RESOURCES

136. *Jean Talon built the first commercial brewery in Quebec City in 1668.*

137. *Molson Brewery is located on its original site in Montreal. It was founded by John Molson in 1786.*

138. *Alexander Keith founded his brewery in Halifax in 1820. Keith's Pale Ale is a popular brew still produced by Keith's today.*

139. *John Labatt founded his brewery in London, Ontario, in 1847.*

140. *The Herchimer & Batkins Brewery was founded at Winnipeg in 1872.*

141. *The Sapperton Brewery opened in New Westminster in 1862.*

142. *The average Yukon resident consumes 142 litres of beer annually. The Canadian average is 83 litres. One-hundred and forty-two litres works out to about 17 cases of 24 bottles.*

143. *Folks from the Northwest Territories drink the least amount of beer per capita – about 66 litres per person per year.*

144. *Seagram's Distillery in Montreal, owned by the Bronfman family.*

145. *The Hiram Walker, Gooderham and Wort's distillery was founded in 1858.*

146. *Elliot Lake.*

147. *Coal.*

148. *Asbestos. Researchers learned that the inhalation of asbestos dust led to diseases, including lung cancer. Asbestos use in Canada was subsequently curtailed in the early 1980's.*

149. *Oil Springs, just south of Sarnia, Ontario, began producing oil in 1857, two years before the first American well. Oil Springs was the first drilled well (as opposed to a dug well) and could therefore access oil located deep underground.*

150. *The town is Leduc, Alberta, and its famous well was Leduc #1.*

151. *Nova Scotia. Over 400 miners have lost their lives in Springhill mines since the 1880's. In 1891, 125 miners were killed in an explosion. In 1956, 39 miners perished in a mine collapse and 88 more were trapped and thought to be lost. The*

"Springhill Miracle" occurred three and a half days later when all 88 were brought to the surface alive. In 1958, a "bump" (what miners call roof collapse) trapped 94 miners, of whom 75 died. This would be the last Springhill disaster; the mines were closed shortly thereafter.

152. Coal – The Glace Bay mine is one of the more interesting mines that one can visit. Its shafts extend away from land out under the Atlantic. At one point visitors are actually under hundreds of metres of rock and hundreds of metres of Atlantic Ocean.

153. Most of the crude oil processed in Montreal and most of the crude processed and sold east of the Ottawa River originates below Lake Maracaibo. Large tankers coming up from Venezuela can't negotiate the St. Lawrence River. Instead they off-load in Portland, Maine. The crude flows north in the Portland Pipeline through Maine, New Hampshire, Vermont, and Quebec, and finally passes under the St. Lawrence and into Montreal East's refineries.

154. D. 80%

155. Many single-industry towns have gone bankrupt because of sudden drops in foreign demand. Towns such as Elliot Lake, Ontario (uranium), Uranium City, Saskatchewan, and Schefferville, Quebec (iron), have each experienced financial ruin and depopulation when foreign markets sought cheaper minerals elsewhere.

156. C. 60

157. D. Alberta – Ontario – B.C. – Quebec

158. A. Petroleum and natural gas
The leading four minerals in the Canadian mining industry are ranked according to value from A to D. Oil and gas account for almost half of all mining revenues. Provinces like Alberta, which are all heavily committed to petroleum and natural gas exploitation, maintain inherent economic advantages. It's little wonder Newfoundland and Nova Scotia hope to cash-in respectively on Hibernia and Sable Island.

159. Gold.

160. Electrical wire and wire products.

161. Northern Labrador, Newfoundland.

162. *The abundance of cheap, plentiful electrical power draws the smelters to Canada and makes them cost effective. Canada's largest supplier of bauxite is Jamaica. Other suppliers are Australia, Guyana, and Guinea.*

163. *Quebec leads by a long shot because Hydro-Quebec offers smelter owners incredibly low electricity prices.*

164. *Arvida is a compound term formed from the first letters of the past president of the Aluminum Company of Canada (now Alcan), Arthur Vining Davies. His company built Canada's first super smelter.*

165. *At mines in the Eastern Townships of Quebec. The center of the strike was at Asbestos and pitted workers against the management of the Johns Manville Company. The struggle lasted 120 days.*

166. *Pierre Trudeau supported the workers and in 1956 edited a collection of pro-labour essays called, La Grève de L'amiante. His future antagonist, René Lévesque, reported on the strike for Radio Canada.*

167. *Winnipeg – a general strike was called by the Winnipeg Trade and Labour Council when negotiations between striking metal and building trades workers and their employers broke off. Over 30,000 workers responded. This was an era of high post-war inflation, low wages, and generally unsafe working conditions. Protests led to violence and two strikers were shot and killed by police. Twenty more strikers were injured.*

168. *The surplus is sold to the adjoining provinces of Ontario and New Brunswick, and to adjoining American states, most notably New York, Vermont, and New Hampshire.*

169. *Yellowknife in the Northwest Territories, and Val d'Or, Quebec. Some streets in these cities are paved with gold ore extracted from nearby mines.*

170. *Gold was discovered in Bonanza Creek, a tributary of the Klondike River, in August of 1896. The creek is near Dawson in west central Yukon. Pierre Berton estimated that as much money was spent by the thousands of prospectors just to get to the Yukon as was actually made from gold rush mining and prospecting.*

171. *True.*

172. *Definitely Newfoundland. St. John's Water Street merchants made fortunes dealing in skins brought back from the ice.*

173. *On drift ice in the Gulf of St. Lawrence.*

174. *On April 2, 1914, sealers from the ship Newfoundland were put out on the drift ice and, through a misunderstanding and with no radio communications, were abandoned on the ice for two days without shelter. Seventy-eight of one hundred and twenty-three sealers died. Many of the survivors were maimed for life. As the Newfoundland's seamen were fighting for their lives, the sealing ship S.S. Southern Cross sank only a few miles away. All 173 sealers aboard her were lost, along with over 17,000 seal pelts.*

175. *Matheson, near Timmins in Northern Ontario, was overrun by a forest fire in 1916. The entire community was destroyed, and 223 people killed.*

176. *The Porcupine Fire of 1911 destroyed three towns: Cochrane, Pottsville and South Porcupine. Seventy-three people died. The second fire occurred in 1922 and is known as the Haileybury Fire. The town of Cobalt was destroyed. Six-thou-sand people were left homeless and forty-four people were killed.*

177. *Insects damage about nine times more forest than fire. Fire destroys about 2.25 million ha of forest per year.*

Connections & Barriers

Bridges

Transportation

Barriers

Rivers

Events

CONNECTIONS & BARRIERS

Connections & Barriers

Bridges

Transportation

Barriers

Rivers

Events

1. Where is the Jacques Cartier bridge located?

2. In which province is the longest railway tunnel?

3. What is the name of Canada's longest bridge?

4. This huge toll bridge links Canada's smallest province (P.E.I.) with the mainland (New Brunswick). What body of water does it cross?

5. What is Canada's second longest bridge?

6. What two cities are connected by the Ambassador Bridge?

7. What cities does the Angus L. Macdonald Bridge connect?

8. What bridge is just north of the Angus Macdonald?

9. Where is the Skyway Bridge found?

10. Where can you find Lions Gate Bridge?

11. Name the town and province in which the longest covered bridge in the world is located.

12. Montreal is centred on an island, but some of its suburbs are found on the mainland. A large number of daily commuters must therefore cross rivers to get downtown. How many bridges connect Montreal to the mainland?

13. What American state would you enter if you crossed the Rainbow Bridge?

14. Where would one find an *Inukshuk?*

15. Where is the oldest lighthouse in Canada?

16. What was the *Underground Railroad?*

17. Where was the Underground Railroad's most northern terminus?

18. Long before the Underground Railway was established, late eighteenth-century Nova Scotia became a terminus for fugitive African-Americans. Why?

19. On which side did most Canadians fight during the American Civil War?

20. Where is the Icefields Parkway and what two major tourist towns does it connect?

21. Where is the Cabot Trail?

22. What towns are located at the distant ends of the Trail?

23. What provinces does the Tete Jeune or Yellowhead Highway run through?

24. At what communities does the highway start and end?

25. Which province is home to the Coquihalla Highway?

26. The Trans-Canada Highway was opened in 1962 and is 7,800 km long. St. John's is its eastern terminus. Where is mile 0 in the West?

27. The Alaska Highway ends at Fairbanks, Alaska, but most of it is found in Canada. Where in Canada does it start?

28. What was the mainstay of the Yukon's transportation system from the 1860's to 1950?

29. Where did the "silk trains" run?

30. Why were these trains in such a rush?

31. What Ontario city acts as a rail centre for eastbound shipments of Western grain?

32. Which were built in Canada first, canals or railroads?

33. The Canadian Pacific Railway runs through which Rocky Mountain passes?

34. The first truly sea-to-sea Canadian railway system was built by...?
   A. Canadian National
   B. Grand Trunk
   C. Ontario Northland
   D. Canadian Pacific

35. Name the Great Lake on which you can travel farthest inland from the Atlantic Ocean. In other words, which lake is farthest west?

36. Which city marks the western end of the St. Lawrence Seaway?

37. Where was the Bricklin automobile built?

38. The Buick was a Canadian car originally developed by a competitor of Henry Ford's. Who was the competitor and where was the Buick built?

39. Which Canadian city had the first subway?

40. Which Canadian city was next to open a subway?

41. Where did "The Bullet" run?

42. Match the city to the Y prefix (flight destination code). These are the three initials on the tag that is attached to our luggage when we check in for a flight.

YOX	Regina
YQR	Toronto
YUL	Gander
YVR	Ottawa
YOW	Vancouver
YYZ	Montreal
YYG	Winnipeg
YNS	Calgary
YHZ	Nemiscau
YYC	Pond Inlet
YHM	Halifax
YGP	Kelowna
YIO	Hamilton
YWG	Charlottetown
YCW	Gaspé

43. What airport handles the most flights and passengers?

44. What is Kelly's Lake?

45. That quintessential Canadian invention, the Ski-Doo, is built where?

46. Who named the Lachine Rapids?

47. Aside from the Lachine Rapids which prevented movement of ships upstream, what two other major physical handicaps did European explorers encounter when sailing ships up the St. Lawrence?

48. Which two countries share the longest border in the world?

49. What is the length of the Canada – United States border?
    A. 6,900 km
    B. 5,900 km
    C. 10,900 km
    D. 8,900 km

50. Name the three closest nations to Canada, excluding the United States?

51. Name the strait separating Cape Breton Island from the mainland.

52. What body of water separates Newfoundland from Quebec and Labrador?

53. Name the body of water separating Greenland from Canada.

54. Where is the Kings Road, and what is its proper name?

55. Which province maintains the largest commercial ferry fleet in North America?

56. How many ferry routes connect Newfoundland to the rest of Canada?

57. How many ferry routes connect Vancouver Island to the mainland?

58. Welland, Trent, and Rideau are all names associated with what type of artificial waterway?

59. What stream runs through Regina?

60. What river runs through Red Deer, Alberta?

61. What is the name of the largest river that runs through Calgary?

62. What is the name of the river that runs through Trois Rivières and Shawinigan, Quebec?

63. What river runs through Fredericton and Saint John, New Brunswick?

64. What river splits Moncton, New Brunswick, and its suburbs?

65. What river divides much of the Chicoutimi-Jonquière area in Quebec?

66. Across what body of water would you be sailing if you were to travel directly from the southern-most tip of the Queen Charlotte Islands to the northern-most point of Vancouver Island?

CONNECTIONS & BARRIERS

67. What is the longest river in Canada, and into what sea does it flow?

68. Into which sea does the Yukon River flow?

69. What potential problem can be associated with long, north-flowing rivers like the Mackenzie?

70. In 1997 a river in Western Canada experienced spring flooding. The result was one of the largest financial disasters in Canadian history. Name the river.

71. The above-mentioned 1997 flood was controversial because of alleged misuse of flood prevention devices that were constructed in the Winnipeg area in the 1950's. What are these devices?

72. At the end of the summer of 1996, there was disastrous flooding along another large Canadian river. Name this Eastern Canadian river.

73. Who was accused of alleged misuse of hydroelectric reservoirs and control dams thereby causing the flooding of the above river?

74. In which Canadian province does the Peace River originate?

75. What large river has its source and mouth in the United States yet most of its flow in Canada?

76. Which provincial capital is located on the North Saskatchewan River?

77. Name the gateway river that allowed European explorers to penetrate far into North America from the east coast?

78. On a canoe trip, you would be travelling upstream during which of the following voyages?
A. Fraser River – from Prince George to Vancouver
B. Assiniboine River – from Brandon to Winnipeg
C. Ottawa River – from Ottawa to Montreal
D. St. Lawrence River – from Montreal to Kingston

79. Newfoundland and Toronto both have rivers that share *what* same name?

80. Niagara Falls is on the Niagara River between Lake Erie and Lake Ontario. What canal allows ships to bypass the falls?

81. *Odd One Out*— Which river doesn't belong and why?
A. Albany
B. Rupert
C. Winisk
D. Churchill
E. Nelson

82. The Ottawa River is to Ottawa as *what* river is to Fredericton?

83. In the eighteenth century, the site where two rivers joined at what is now Winnipeg made a good location for a fur traders' fort. Name the two rivers.

84. *ODD ONE OUT* — Pick the river that doesn't belong and explain why.
A. Peace River
B. Mackenzie River
C. Taltson River
D. Hay River
E. Yellowknife River

85. Name the Canadian waterfall which has the greatest flow of water.

86. What natural event shut down the Niagara Falls on March 29, 1848?

87. Did nature ever shut down the falls again?

88. The last time water ceased to flow over the Falls was in 1969 when the American Falls were shut down. Why?

89. What Canadian waterfall on the South Nahanni River is nearly twice the height of the Horseshoe Falls?

90. Which island in the Niagara River separates the Canadian and American Falls?

91. What province and what state border the Niagara River?

92. Which bridge took 17 years to complete, collapsed twice during its construction, and took the lives of 88 workers?

93. This 1958 explosion is still considered the largest non-nuclear man-made explosion in the world. It was detonated underwater in a successful effort to remove the notorious shipping obstacle, Ripple Rock. Where was Ripple Rock?

94. What Victoria, B.C., bridge gave way under the weight of an overcrowded streetcar in 1896, killing 55 people?

95. Two years after the sinking of the *Titanic*, a marine disaster occurred just east of Rimouski on the St. Lawrence River. What was the name of the ill-fated ship which sank?

96. What caused the disaster?

97. What was the Victoria Day Disaster and where and when did it occur?

98. What ferry was torpedoed during World War II in the Cabot Strait while travelling between Channel-Port aux Basques, Newfoundland, and North Sydney, Nova Scotia.

Connections & Barriers

Answers

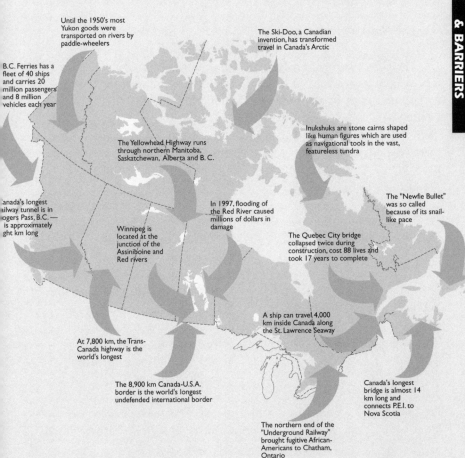

Until the 1950's most Yukon goods were transported on rivers by paddle-wheelers

The Ski-Doo, a Canadian invention, has transformed travel in Canada's Arctic

B.C. Ferries has a fleet of 40 ships and carries 20 million passengers and 8 million vehicles each year

Inukshuks are stone cairns shaped like human figures which are used as navigational tools in the vast, featureless tundra

The Yellowhead Highway runs through northern Manitoba, Saskatchewan, Alberta and B. C.

The "Newfie Bullet" was so called because of its snail-like pace

In 1997, flooding of the Red River caused millions of dollars in damage

anada's longest ailway tunnel is in ogers Pass, B.C. — is approximately ght km long

Winnipeg is located at the junction of the Assiniboine and Red rivers

The Quebec City bridge collapsed twice during construction, cost 88 lives and took 17 years to complete

A ship can travel 4,000 km inside Canada along the St. Lawrence Seaway

At 7,800 km, the Trans-Canada highway is the world's longest

The 8,900 km Canada-U.S.A. border is the world's longest undefended international border

Canada's longest bridge is almost 14 km long and connects P.E.I. to Nova Scotia

The northern end of the "Underground Railway" brought fugitive African-Americans to Chatham, Ontario

1. Montreal. The Jacques Cartier is one of three downtown bridges linking the island city to the mainland.

2. It is located at Rogers Pass in B.C. and is approximately eight km long.

3. The Confederation Bridge, which is 14 km long and joins P.E.I. to mainland Nova Scotia.

4. The Northumberland Strait.

5. The Pierre Laporte Bridge is 670 m long and connects Quebec City to the south shore of the St. Lawrence. This bridge was named after Quebec Minister of Labour, Pierre Laporte, who was assassinated by F.L.Q. terrorists during the October Crisis of 1970.

6. Windsor, Ontario, and Detroit, Michigan.

7. Halifax and Dartmouth.

8. The Murray McKay Bridge.

9. Burlington/Hamilton, Ontario.

10. Vancouver – it spans the First Narrows and connects North Vancouver to Vancouver.

11. Hartland, New Brunswick, has the longest covered bridge. This 391 m long bridge is made of wood and crosses the St. John River.

12. There are 15 auto bridges (one is a combination bridge-tunnel) leading to Montreal. In addition, there are 6 railway lines connecting the island.

13. The Rainbow Bridge spans the Niagara River between Ontario and New York state.

14. An Inukshuk is a stone cairn shaped like a human. Inukshuks are found in the Arctic where they are used as navigational landmarks.

15. Fisgard Lighthouse is located a few kilometres from Victoria, B.C. It was built in 1873.

16. The Underground Railroad had little to do with transportation in the conventional sense. Rather, it was a system of safe houses set up in the United States to shelter escaped slaves from slave hunters and help them flee the South. Fugitives travelled northward from safe house to safe house. The Railroad operated between 1840 and 1860. Thirty-thousand escaped slaves are estimated to have made their way to Canada through the Underground Railroad.

17. *The Leamington/Chatham area in Southern Ontario, home to what was the largest nucleus of African-Americans in Canada at the time.*

18. *Although slavery was legal in Nova Scotia, fugitive African-Americans came to Nova Scotia because they had been promised free land if they helped fight for the British in the American Revolution. They came to Halifax to collect their land despite the fact that the British had lost. Some received land in Nova Scotia but most were given free land in Africa's Sierra Leone. It is estimated that in 1792, 15 ships left Halifax for Sierra Leone taking over 1,200 would-be land owners to Africa.*

19. *Fifty-thousand Canadians fought for the North in the American Civil War. Fewer than 1,000 Canadians fought for the South.*

20. *The Parkway runs through the Rocky Mountains in Alberta north from Banff, through Lake Louise, passes the Columbia Icefields, spectacular mountains, rushing streams and waterfalls and ends at Jasper 225 km later.*

21. *The Cabot Trail is a 296 km highway noted for its rugged mountain and coastal scenery. The trail attracts hundreds of thousands of tourists to Cape Breton, Nova Scotia, every year.*

22. *The Trail is a closed circular loop so there are no "ends." The main town on the Trail is Baddeck where Alexander Graham Bell kept his summer home. There is much discussion as to whether the drive should be done in a clockwise or counter-clockwise fashion. In this day of effective sunglasses, most people don't mind driving "into" the sun. However, some people do express nervousness when driving along the cliffside of seemingly interminable drops to the ocean far below.*

23. *Highway 16 or The Yellowhead is sometimes referred to as the Northern Trans-Canada Highway. It runs through Manitoba, Saskatchewan, Alberta, and B.C.*

24. *It starts just west of Portage la Prairie, Manitoba, and ends 2,800 km later at Prince Rupert, B.C., on the Pacific Ocean.*

25. *B.C. – this highway runs north/south between the Fraser River and Okanagan Lake.*

26. *The Trans-Canada Highway ends (or begins) at Victoria, B.C., making it the longest paved highway in the world.*

27. *This highway, built to move goods and personnel during World War II, starts at Dawson Creek, B.C.*

28. *Riverboats, or sternwheelers, transported almost all people and goods in and out of the Yukon before the spread of modern ground and air transportation.*

29. *These express trains ran from Vancouver to Montreal between the early 1900's and 1930's and carried valuable Oriental silk to Eastern merchants.*

30. *The silk on the trains had a value of about $6 million per shipment. The merchandise was insured by the hour. Fast delivery meant lower insurance costs.*

31. *Thunder Bay.*

32. *Canals.*

33. *Kicking Horse Pass and Rogers Pass.*

34. *D. Canadian Pacific*

35. *Lake Superior.*

36. *The Seaway, which opened in 1959, stretches from Anticosti Island in the Gulf of St. Lawrence to Duluth, Minnesota, at its most western end, a distance of almost 4,000 km. Canadians think of Thunder Bay as the western terminus, but Duluth is farther inland, or westward.*

37. *For two years, this luxury two-seat sportscar was built in Saint John, a fitting location because Malcolm Bricklin had persuaded the New Brunswick government to help bankroll his automotive ambitions.*

38. *Sam McLaughlin introduced his Buick in Oshawa, Ontario, the same year that Ford introduced his Model T.*

39. *Toronto opened its subway in 1954.*

40. *Montreal opened the Métro in 1966.*

41. *This slow, local, CNR train ran for many years across Newfoundland until it was replaced by buses. The train was ironically nicknamed "The Bullet" because it was notoriously slow.*

42. YOX – Gander, Nfld.
YUL – Montreal, Que.
YOW – Ottawa, Ont.
YYZ – Toronto, Ont.
YVR – Vancouver, B.C.
YNS – Nemiscau, Que.
YHZ – Halifax, N.S.
YYC – Calgary, Alta.
YHM – Hamilton, Ont.
YGP – Gaspé, Que.
YIO – Pond Inlet, N.W.T.
YCW – Kelowna, B.C.
YYG – Charlottetown,
    P.E.I.
YQR – Regina, Sask.
YWG – Winnipeg, Man.

43. Toronto's Pearson International Airport leads all Canadian airports in both categories.

44. It is the original name of what is now known as the Halifax International Airport.

45. It is built near Sherbrooke, Quebec, in Valcourt, the town in which Armand Bombardier lived when he invented the Ski-Doo.

46. Jacques Cartier is thought by many to have named this obstacle on his westward travels. But it was really Cavelier de La Salle who named them because they blocked his path to China – or La Chine as it is known in French.

47. They were sailing upstream against the current and largely into the prevailing westerly winds.

48. Canada and the United States.

49. D. 8,900 km. The length of the Alaska – Yukon border is 2,500 km; our southern border constitutes the remaining 6,400 km.

50. Russia (over the Arctic Ocean), Denmark (because of its continuing relationship with Greenland), and France (because of its territorial ownership of St. Pierre and Miquelon).

51. Strait of Canso.

52. Strait of Belle Isle.

53. Davis Strait.

54. Open for business in 1734, Le Chemin du Roi links Montreal and Quebec, along the northern shore of the St. Lawrence River.

55. B.C. owns and operates a 40-ship fleet. These ferries carry 21 million passengers and 8 million vehicles every year.

56. There are two routes – both leading to North Sydney on Cape Breton Island. From

Newfoundland, one may board the ferry at Port-aux-Basques on the western part of the island for a six-hour trip. From the eastern part of the island, one may leave from Argentia for a twenty-hour trip to the mainland.

57. Vancouver Island has four Canadian services run by B.C. Ferries – one from Comox on the Island to Powell River, and three more southern routes, one from Horseshoe Bay north of Vancouver to Nanaimo on Vancouver Island; one from Tsawwassen, south of Vancouver, to Nanaimo; and another from Tsawwassen to Swartz Bay, close to Victoria. In addition, there are two ferries to the United States: one from Victoria to Seattle, Washington, and another to Port Angeles, Washington, near Olympic National Park. All these ferries accommodate both automobiles and people.

58. Canals.

59. Wascana Creek.

60. Red Deer River.

61. Bow River.

62. St. Maurice River.

63. Saint John River.

64. Petitcodiac River.

65. Saguenay River.

66. Queen Charlotte Sound.

67. The Mackenzie River. It flows into the Beaufort Sea, part of the Arctic Ocean.

68. The Bering Sea.

69. When the source, as in the case of the Mackenzie, is located south of the river's mouth, it may experience spring break-up before the colder frozen areas downstream. The southern waters are forced to flow over their banks.

70. The Red River in Manitoba.

71. Floodways – huge channels over 50 km in length excavated to accommodate extra flood water. The floodways are designed to guide flood water from the Red River around Winnipeg along the east side of the city.

72. The Saguenay.

73. Hydro Quebec which controlled reservoir levels.

74. B.C.

75. The Columbia River.

76. Edmonton.

77. St. Lawrence River.

78. D. St. Lawrence River – from Montreal to Kingston

79. Humber River.

80. The Welland Canal was first opened in 1829 and has undergone a number of improvements to keep it modern and efficient.

81. All flow into James Bay and Hudson Bay in an easterly direction with the exception of the Rupert which flows westward.

82. The St. John River.

83. The Assiniboine River and the Red River. The site was called The Forks.

84. All these rivers empty into Great Slave Lake except the Mackenzie which flows out of the Lake. Therefore, the answer is the Mackenzie.

85. The Horseshoe or the International Falls of Niagara Falls, which is 54 km high and has a flow of 155 million litres per minute.

86. The Niagara River stopped flowing because ice lodged in the river at the point where it flows out from Lake Erie, causing the falls to run dry. The blockage lasted until March 31, 1848.

87. Yes, six times in total. The Falls were shut off (sometimes more than once in a year) in 1909, 1936, and 1947.

88. The U.S. Army Corps of Engineers built a coffer dam from the New York shore to Goat Island and diverted the entire flow of water over the Horseshoe or International Falls. This was done so the Army could repair the crumbling edge of the American Falls and clear debris at the base. They wisely chose to do little because it is the debris that make the American Falls so unique, especially when illuminated by spotlights from the evening light show.

89. Virginia Falls in the southwestern Northwest Territories is 90 m in height.

90. Goat Island.

91. Ontario on the Canadian side and New York on the American.

92. *The Quebec City Bridge, which collapsed in 1907 due to faulty engineering calculations. Seventy-five workers died in the accident. A new bridge was built using the same design, but a construction accident in 1916 claimed 13 more lives.*

93. *It was located south of Victoria, in the Strait of Juan de Fuca, between Vancouver Island and the Olympic Peninsula in the state of Washington.*

94. *The Point Ellice Bridge.*

95. *The* Empress of Ireland *sank in 1914. Four hundred and sixty-five people escaped, but over 1,000 lost their lives (almost as many as in the* Titanic *tragedy).*

96. *The Norwegian coal carrier,* Norstad, *rammed the Canadian Pacific liner which sank in 14 minutes. This was the worst marine disaster in Canadian waters.*

97. *The accident occurred on May 24, 1881 (Victoria Day), and involved the capsizing and subsequent deck collapse of the pleasure ferry,* Victoria. *The ferry keeled over on the Thames River, Ontario. One hundred and eighty-one lives were lost.*

98. *The* Caribou – *137 people lost their lives.*

Locations

Districts
Latitude & Longitude
Place Names

1. Mount Royal is a small mountain, and also a suburb in Montreal. What large Prairie city also includes a suburb named Mount Royal?

2. Where is Gastown?

3. Where is Cabbagetown?

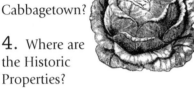

4. Where are the Historic Properties?

5. Where is Lower Town?

6. Where is Shaughnessy?

7. Where is Old Strathcona?

8. Where is Bore View Park?

9. Where is the Forestry Farm and Animal Park?

10. In what province is Cariboo Country?

11. Where is Curling?

12. Where is Byward Market?

13. Where is the Miramichi River?

14. Where is the Interlake area?

15. Near what city is Butchart Gardens?

16. Where is the International Peace Garden?

17. Where is Rockcliffe?

18. Where is Point Pleasant Park?

19. Where is Glenmore Reservoir?

20. Where is the National Exhibit and Centre for Indian Art?

21. Where is Province House National Historic Site?

22. Where are the Plains of Abraham?

23. Where is the Bridle Path?

24. Where is Sea Island?

25. Where is the Loyalist Burial Ground?

26. Where is Ramsay Lake?

27. Where is the RCMP Museum and Barracks?

28. Where is Quidi Vidi Lake?

29. Where is the Ukrainian Museum of Canada?

30. Where is Moose Factory ?

31. Where are the Spirit Sands?

32. Where is the Carcross Desert?

33. Name the line of latitude that forms part of the world's longest undefended international border?

34. Ontario shares land borders with two provinces and three states. Name them.

35. What American border states are found south of the 49th parallel?

36. What other states border Canada?

37. What three states border the Great Lakes but do not contact Canadian soil?

38. Since we're dealing with American states, how many of the 28 states located wholly or partially north of our most southern point can you name? (Note: Canada's most southern

point is a tiny island in Lake Erie called Middle Island located south of Point Pelee. Middle Island is just south of 42° north latitude.)

39. What is so odd about Point Roberts in the state of Washington?

40. What is odd about the Northwest Angle in Minnesota?

41. The Alaska Panhandle occupies half of what one might think should be B.C.'s coastline. Why is the Panhandle part of Alaska?

42. Name the city located on the 105° east longitude line.

43. What is Canada's most western spot?

44. What part of Canada is first to receive morning

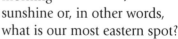

sunshine or, in other words, what is our most eastern spot?

45. What is Canada's most northern location?

46. What is Canada's north-ernmost permanent settlement?

47. What is Canada's highest point?

48. What is Canada's lowest point?

49. Where is Canada's *geographical centre*?

50. Is Canada taller north to south or wider east to west?

51. What country extends farther west – Canada or the United States?

52. To what does the term North of 60 refer?

53. What two provinces extend north of the 60° north latitude line?

54. What are the Torngats?

55. When was Newfoundland granted the remainder of the territory that we now know as Labrador?

56. What is expressed by the statement, "Toronto is closer to the Atlantic than to the Pacific."
    A. Absolute location
    B. Co-ordinates
    C. Relative location
    D. Vectors

57. Toronto is how many times closer to the Atlantic than to the Pacific?
    A. 4
    B. 3
    C. 2
    D. 5

58. What Canadian engineer developed the Intercolonial Railway routes from Quebec to the Maritimes, and the Transcontinental Railway from Montreal to the Pacific.

59. In 1884, an international conference on time zone standardization was held in Washington, D.C., which was largely the consequence of Sir Sandford Fleming's work. One result of this conference was a world divided into twenty-four Standard Time Zones whereby all locations within each 15° zone kept the same standard time. How many time zones are in Canada?

60. In which time zone is Winnipeg?

61. Are clocks in Halifax normally ahead or behind those in Winnipeg?

62. The International Date Line runs through the Pacific Ocean and what other world ocean?

63. In what direction would you travel if you left Medicine Hat for Moose Jaw?

64. Driving from Edmonton to Calgary would take you in what general direction?

65. An airplane leaves Saskatoon for Regina. In what direction will the pilot fly the plane?

66. If you were driving from Regina to Moose Jaw, in what cardinal direction would you be travelling?

67. If you drove due south from Swift Current, Saskatchewan, which American state would you enter first?

68. Does any province touch more than two other provinces?

69. How many provinces touch only one other province?

70. Is St. John's, Newfoundland, closer to England or to Canada's West Coast?

71. If you were sailing west from the Hibernia oil field, which province would you arrive at first?

72. If you were on a ship travelling up the St. Lawrence River, which Great Lake would you encounter first?

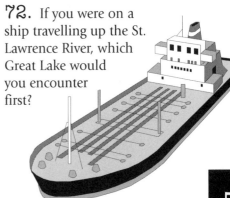

73. Which degree of latitude marks the boundary between Saskatchewan and the Northwest Territories?

74. 110° west longitude separates Saskatchewan and which province?

75. Does the 102° west longitude line mark the border between Saskatchewan and Manitoba?

76. The North Magnetic Pole is located in Canada. Where is it found?

77. What Canadian city sits right atop a provincial boundary?

78. *Kingston* is a small but diverse city located in eastern Ontario. It was originally a small French fort called Fort Cataraqui founded in 1673. Kingston is also the name of the capital of what Caribbean nation?

79. *Halifax* is the largest city in the Atlantic provinces. It was named after a small city in another country. In what country is the original Halifax?

80. *Sydney* is the second largest city in Nova Scotia. There is another Sydney located in the southern hemisphere that was founded, like its Canadian counterpart, in the 1780's. What country is it in?

81. *Waterloo* is a city in southwestern Ontario that is home to two large universities. It is also the name of a city where Napoleon Bonaparte met his final defeat. In what country did Napoleon meet his Waterloo?

82. *Edmonton*, Alberta, is home to the CFL's Eskimos and the NHL's Oilers. It was named after the suburb of a large city in Europe. Where is the original Edmonton?

83. *Kimberley* is a small city in B.C. noted for its zinc, lead, and silver deposits. Another Kimberley is also noted for its mining. This city produces the most diamonds in the world. In what country is this non-Canadian Kimberley found?

84. *Kitchener* is a city in southwestern Ontario. Its original name was changed during World War I for patriotic reasons. The original name was considered "suspicious," because it was also the name of an enemy German city. What was Kitchener's original name?

85. *Lunenburg*, Nova Scotia, was recently chosen by UNESCO as a World Heritage Site. It was settled primarily by people from a Lunenburg located elsewhere in the world. What is the name of the country that many old-time Lunenburgers claim as their ancestral home?

86. *Selkirk* is a farming/suburban town north of Winnipeg. It was named after a European noble who was instrumental in settling Manitoba. From what country did this noble originate?

87. *Montréal* is the largest city in Quebec, four times the size of the next largest city, Quebec City. Montreal was originally called

Hochelaga. Its name was changed to one borrowed from the home of some of its original settlers. From what country did these settlers originate?

88. *Banff* is the name of perhaps the most "touristy" town in Canada. It is located in the Rocky Mountains of Alberta. It is also in the most popular National Park in Canada – judging from the number of visitors it receives. From what country does the name Banff spring?

89. *Mount Cook* is a mountain peak in the St. Elias Mountains, on the Yukon-Alaska border. There is another Mount Cook in the Southern Alps. Where are the Southern Alps?

90. *Marathon* is a small community on the north shore of Lake Superior. The long drive between Sault Ste. Marie and Thunder Bay (Marathon lies in the middle) would remind anyone of a marathon. The first marathon was raced in what country?

91. *Newcastle* is a town on the Miramichi River in New Brunswick. It is also a town on Lake Ontario, near Oshawa, Ontario, and the name of communities in Australia, Northern Ireland, South Africa, the United States, and St. Kitt's and Nevis in the Caribbean. Where is the original Newcastle?

92. *Blenheim* is a small community in Ontario midway between London and Windsor. There is also a Blenheim in the southern hemisphere in another old British colony. This small country of a few million people is located on two main islands. In what country is this Blenheim?

Identify the geographic features associated with the following names:

93. Petitcodiac

94. Coppermine

95. Ungava

96. Akimiski

97. Kananaskis

98. Grasslands

99. Cedar

100. St. Elias

101. Fogo

102. Honguedo

103. Rossignol

104. Malpeque

105. Magnetic Hill

106. Conception

107. Texada

108. What do *A Lake* in New Brunswick and *Lac Y* in Quebec have in common?

109. It shouldn't be too difficult to determine why Pekwachnamaykoskwaskway-pinwanik Lake would be included in a question here?

110. What is the capital of Nunavut?

111. What was it formerly called?

112. What is the difference between the Atlantic provinces and the Maritimes or Maritime provinces?

113. What is the difference between the Prairie provinces and Western provinces?

The next questions are short and snappy. We'll give the historical name. You provide the current name.

114. Bytown

115. Statacona

116. Pile O' Bones

117. Abegweit

118. Île Royale

119. Sainte Anne

120. Red River Colony

121. Île St. Jean

122. Fort Garry

123. Camosack

124. Port Arthur

125. Lower Canada

126. Which city is called "Steel City"?

127. Which city is called "The Hartford of Canada"?

128. Which city is called the "Warden of the North"?

129. Which city is called the "City of Bridges"?

130. Which province is the "Garden of the Gulf"?

131. Which province is "Wild Rose Country"?

132. The arctic poppy is the representative flower of what territory?

133. The Stellar's jay is its provincial bird; the dogwood its provincial flower. Name this province.

134. The red spruce is its provincial tree; the mayflower its provincial flower. Name this province.

135. "The small under the protection of the great" is what province's motto?

136. "Glorious and free" is what province's motto?

137. "Hope Restored" is what province's motto?

138. What was the name of the huge French fortress on the east coast of North America?

139. An important War of 1812 event was the capture of Fort York. Where was Fort York?

140. Who established Fort St. James on B.C.'s Stewart Lake?

In the following question, we'll give you the name of the old military fort, you provide the name of the province associated with each:

141. Fort Beauséjour

142. Fort Prince of Wales

143. Fort Amherst

144. Fort Chambly

145. Fort Walsh

146. Fort Nashwaak

147. Fort Edward

148. Fort St. Marie

149. Fort Battleford

150. Fort Carlton

151. Fort Cotêau

152. Fort Mississauga

153. Fort Whoop-Up

154. Fort Normandeau

155. Fort Macleod

156. Fort Malden

157. Fort Lennox

158. Fort Wellington

159. Fort Qu'appelle

160. Quidi Vidi Battery

161. Fort Anne

162. George Island

163. Fort George

164. Fort Henry

165. Fort Langley

Connect the following area codes with the corresponding provinces or regions:

604	Yukon/N.W.T./Nunavut
403	New Brunswick
306	Montreal
204	Southern Alberta
613	Vancouver Region
514	Newfoundland
506	Saskatchewan
902	Ottawa Valley
709	Nova Scotia & P.E.I.
867	Manitoba

Locations

Answers

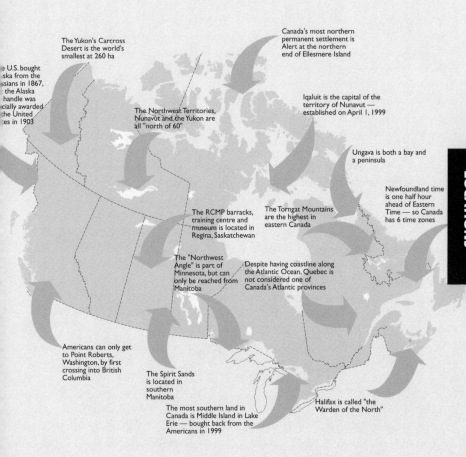

The Yukon's Carcross Desert is the world's smallest at 260 ha

Canada's most northern permanent settlement is Alert at the northern end of Ellesmere Island

e U.S. bought ska from the ssians in 1867, the Alaska handle was cially awarded the United es in 1903

Iqaluit is the capital of the territory of Nunavut — established on April 1, 1999

The Northwest Territories, Nunavut and the Yukon are all "north of 60"

Ungava is both a bay and a peninsula

Newfoundland time is one half hour ahead of Eastern Time — so Canada has 6 time zones

The RCMP barracks, training centre and museum is located in Regina, Saskatchewan

The Torngat Mountains are the highest in eastern Canada

The "Northwest Angle" is part of Minnesota, but can only be reached from Manitoba

Despite having coastline along the Atlantic Ocean, Quebec is not considered one of Canada's Atlantic provinces

Americans can only get to Point Roberts, Washington, by first crossing into British Columbia

The Spirit Sands is located in southern Manitoba

The most southern land in Canada is Middle Island in Lake Erie — bought back from the Americans in 1999

Halifax is called "the Warden of the North"

1. Calgary, Alberta.

2. Vancouver.

3. Toronto.

4. Halifax.

5. Quebec City.

6. Vancouver.

7. Edmonton.

8. Moncton.

9. Saskatoon.

10. B.C.

11. It is a suburb of Cornerbrook, Newfoundland.

12. Ottawa.

13. New Brunswick.

14. Between Lake Winnipeg and Lake Manitoba, Manitoba.

15. Victoria.

16. The 946 ha International Peace Garden spans the Canada-U.S. border at Boissevain, Manitoba/Dunseith, North Dakota. Established in 1932, the Garden celebrates the peaceful relationship between Canada and the U.S.

17. Ottawa.

18. Halifax.

19. Calgary.

20. Thunder Bay, Ontario.

21. Charlottetown.

22. Quebec City.

23. Toronto.

24. Vancouver.

25. Saint John, New Brunswick.

26. Sudbury, Ontario.

27. Regina.

28. St. John's.

29. Regina.

30. James Bay.

31. Southern Manitoba – this group of shifting sand dunes is also known as the Carberry Desert.

32. The Carcross Desert, in southern Yukon, is the world's smallest desert at 260 ha.

33. 49° north latitude forms the border between Manitoba,

Saskatchewan, Alberta, mainland B.C. and the United States to the south.

34. The provinces are Manitoba to the west and Quebec to the east. The American states from west to east are Minnesota, Michigan, and New York.

35. Minnesota, North Dakota, Montana, Idaho and Washington.

36. From west to east the remaining states that touch Canadian soil are: Michigan, New York, Vermont, New Hampshire, Maine, and back to the northwest touching the Yukon is Alaska.

37. Wisconsin, Ohio, and Pennsylvania.

38. The 28 states wholly or partially north of Middle Island are: Alaska, Washington, Oregon, California, Idaho, Nevada, Montana, Wyoming, Utah, North Dakota, South Dakota, Nebraska, Minnesota, Iowa, Wisconsin, Illinois, Indiana, Michigan, Ohio, Pennsylvania, New York, New Jersey, Connecticut, Rhode Island, Massachusetts, Vermont, New Hampshire, and Maine. Middle Island was purchased by the Nature Conservancy of Canada for just over $1.3 million on July

28th, 1999, and is home to over 35 rare species of plants and animals.

39. Washingtonians travelling by car can only visit this part of their state by first crossing the border into Canada. This haven of cheap liquor and gasoline is located south of Vancouver and occupies the southern tip of a peninsula anchored entirely in Canada. Point Roberts, however, is south of the 49° latitude line and is therefore part of Washington.

40. A similar border oddity: Minnesota residents can only visit this part of their state by first entering Canada.

41. The United States purchased Alaska from Russia in 1867. Americans believed the boundary included the Panhandle as defined by the Anglo-Russian treaty of 1825. Canadians thought otherwise. An international tribunal of 3 American jurists, 2 Canadian, and one British decided the matter in 1903. The United States was awarded the Panhandle because the British jurist sided with the Americans.

42. Regina.

43. The 141° west longitude line marking the border between the Yukon and Alaska.

44. Cape Spear in Newfoundland at about 53° west longitude. There is a lighthouse and, as in much of Newfoundland, a spectacular view of sea and coast.

45. Cape Columbia at 84° north latitude marks Canada's most northern location, a mere 670 km from the North Pole.

46. Alert, Northwest Territories, located at the northern end of Ellesmere Island.

47. Mount Logan in the Yukon with an elevation of 5,950 m.

48. This is a bit of a trick question. Canada has no land below sea level. Its coastline, which is the longest in the world, is its lowest point at 0 metres above sea level.

49. Geographical centre is an interesting term that geographers take to mean different things. A human or social geographer, for instance, will determine the geographic centre through a study of population distribution. In Canada, the centre would be drawn to the south of the country and certainly to the east. As

Canada's population grows, this centre is gradually moving westward but not so much towards the north. Canada's physical centre, or what geographers call a "balance centre" or "balance point" is much farther north. Imagine an inflexible cut-out map of Canada sitting atop a sharp point. Where would this shape balance? It would balance near Eskimo Point in Nunavut on the western shore of Hudson Bay at approximately 94° west longitude and 62° north latitude. This is Canada's Geographical Centre.

50. Most Canadians know that the country is wider than it is tall. Almost every map of the complete country is landscape style rather than portrait style thereby emphasizing the country's width. However, Canada is close to being as tall as it is wide. Our most southern point is about 42° north latitude and our most northern is 84° north latitude. At 111 km per degree of latitude, the difference, 42°, yields a total north-south distance of 4,662 km. Similarly, our east-west distance is 89° from 141° west on the Yukon boundary to 52° east near St. John's. However, a degree of longitude varies in length and this is the key to understanding this answer. If we take the 50° latitude line to measure a degree of longitude, it yields a length of about 50 km per degree. The product of 89° multiplied by 50 km is 4,450. So we can see from both north-south and east-west distances, Canada is almost square. Most Canadians do not know that our most western point is the boundary with Alaska. They think the west coast of Vancouver Island is our most western point but its latitude is 128° west. The Queen Charlotte Islands are only at 133° west. Both, however, fall far short of the 141° Alaskan boundary mark.

51. Canada's most western location is the Yukon-Alaska boundary at 141° west, while the U.S.'s most western location is Attu Island at 173° east longitude.

52. Canadians use "North of 60" to refer to the lands north of the 60th parallel of latitude. The Yukon, Northwest Territories, and Nunavut are all "North of 60."

53. Quebec and Newfoundland, where Labrador touches Quebec at the Torngat Mountains.

54. The Torngats are the Atlantic provinces' highest mountain peaks, located on the Quebec-Labrador border.

55. In 1927, the Imperial Privy Council in London, England, granted Newfoundland the present territory of Labrador.

56. C. Relative location

57. A. 4

58. Sir Sandford Fleming.

59. There are six time zones in Canada. Newfoundland, however, has opted to differ from Atlantic time by half an hour. This makes for a four-and-a-half hour total time difference from one side of the country to the other.

60. Central Time Zone.

61. Ahead.

62. Arctic Ocean.

63. East.

64. South.

65. Southeast.

66. West. Moose Jaw and Regina are on the Trans-Canada Highway 60 km apart.

67. Montana.

68. Only Quebec touches three: Ontario to the west, New Brunswick and Labrador, Newfoundland, to the east.

69. *Just two on the mainland: B.C. and Nova Scotia. P.E.I. and Newfoundland (sans Labrador), touch none.*

70. *This one's not even close. Newfoundland is about 3,200 km from London yet well over 7,000 km from Victoria.*

71. *Newfoundland.*

72. *Lake Ontario.*

73. *60° north latitude.*

74. *It's the line that forms the border between Saskatchewan and Alberta.*

75. *Yes and no. It does in the northern (unsettled) third of Saskatchewan where the line runs straight. However, a large scale map of Saskatchewan shows that the southern part of the eastern boundary is made up of a series of 24 mile or 38 km long steps or notches. These steps represent the corrections made to incorporate the original land divisions of the Prairie Township system.*

76. *It is presently on Ellef Ringnes Island in the Arctic but continues to move constantly in a northwesterly direction at a rate of less than 10 km annually.*

77. *Lloydminster on the Saskatchewan-Alberta border, about 225 km east of Edmonton.*

78. *Jamaica.*

79. *England.*

80. *Australia.*

81. *Belgium.*

82. *England.*

83. *South Africa.*

84. *Berlin.*

85. *Germany.*

86. *Scotland.*

87. *France.*

88. *Scotland.*

89. *New Zealand.*

90. *Greece – Phidippides ran from Marathon to warn the Greeks of a Persian invasion.*

91. *England.*

92. *New Zealand.*

93. *A river in New Brunswick.*

94. *A village in Nunavut. It is now called Kugluktuk.*

95. *A bay in northern Quebec.*

96. *An island in James Bay between Ontario and Quebec.*

97. *A village in the midst of a recreation/wilderness area in Alberta.*

98. *A national park in Southern Saskatchewan.*

99. *A lake in Manitoba and part of Lake Agassiz, the huge inland sea that flooded much of the Prairies after the last period of continental glaciation.*

100. *Canada's second highest mountain, located in the Yukon.*

101. *An island off the north coast of Newfoundland.*

102. *A strait between Anticosti Island and the Gaspé Peninsula at the mouth of the St. Lawrence River in Quebec.*

103. *A lake in western Nova Scotia.*

104. *A bay on the north shore of P.E.I.*

105. *A tourist attraction in Moncton, New Brunswick. Automobiles are seemingly drawn uphill by a mysterious* force. *It is actually an optical illusion, but a very realistic one.*

106. *A bay on the southeast coast of Newfoundland.*

107. *An island off the west coast of B.C. exploited for its limestone and timber.*

108. *They're both lakes which share the shortest names of any place in Canada.*

109. *At thirty-one letters, it's Canada's longest place name.*

110. *Iqaluit was chosen as the capital by the residents of Canada's newest territory.*

111. *Frobisher Bay.*

112. *Basically, the difference is Newfoundland. The Maritime provinces have traditionally included New Brunswick, Nova Scotia, and P.E.I. When Newfoundland entered Confederation in 1949, an all-inclusive term for the four provinces was required. The term Atlantic provinces was the result. Since Quebec also has an eastern sea coast along the Gaspé and St. Lawrence, should it be considered an Atlantic province, too?*

113. B.C. is not a Prairie province. "The Prairies" refers to the southern grassland area east of the Rockies and the term Prairie provinces refers to the provinces of Alberta, Saskatchewan and Manitoba. It seems that in writing about Western Canada, there are times when one needs to include and/or exclude B.C. from provincial groupings which makes these separate terms useful.

114. Ottawa.

115. Quebec City.

116. Regina.

117. P.E.I.

118. Cape Breton.

119. Fredericton.

120. Manitoba.

121. P.E.I.

122. Winnipeg.

123. Victoria.

124. Thunder Bay.

125. Quebec.

126. Hamilton.

127. London.

128. Halifax.

129. Saskatoon

130. P.E.I.

131. Alberta.

132. Nunavut.

133. B.C.

134. Nova Scotia.

135. P.E.I.

136. Manitoba.

137. New Brunswick.

138. Louisbourg.

139. Fort York sat on the site of present day Toronto. It was captured by the Americans after a short battle in 1813.

140. Simon Fraser established this Hudson's Bay Company trading post in 1806. Located 160 km northwest of Prince George, Fort St. James is home to the largest group of wooden fur trade buildings in Canada.

141. Fort Beauséjour – New Brunswick

142. *Fort Prince of Wales – Manitoba*

143. *Fort Amherst – P.E.I.*

144. *Fort Chambly – Quebec*

145. *Fort Walsh – Saskatchewan*

146. *Fort Nashwaak – New Brunswick*

147. *Fort Edward – Nova Scotia*

148. *Fort St. Marie – Ontario*

149. *Fort Battleford – Saskatchewan*

150. *Fort Carlton – Saskatchewan*

151. *Fort Cotêau – Quebec*

152. *Fort Mississauga – Ontario*

153. *Fort Whoop-Up – Alberta*

154. *Fort Normandeau – Alberta*

155. *Fort Macleod – Alberta*

156. *Fort Malden – Ontario*

157. *Fort Lennox – Quebec*

158. *Fort Wellington – Ontario*

159. *Fort Qu'appelle – Saskatchewan*

160. *Quidi Vidi Battery – Newfoundland*

161. *Fort Anne – Nova Scotia*

162. *George Island – Nova Scotia*

163. *Fort George – Ontario*

164. *Fort Henry – Ontario*

165. *Fort Langley – B.C.*

604 – Vancouver Region
403 – Southern Alberta
306 – Saskatchewan
204 – Manitoba
613 – Ottawa Valley
514 – Montreal
506 – New Brunswick
902 – Nova Scotia & P.E.I.
709 – Newfoundland
867 – Yukon/N.W.T./Nunavut

LOCATIONS

Communities

Cities
Events & Exhibitions
Legends
Tourist Attractions
Nationhood
Icons
Universities
Events

Questions 1 to 5 concern twin cities. Twins are adjacent to one another and often maintain economic ties. Name the twin of each of the following.

1. Waterloo, Ontario

2. Halifax, Nova Scotia

3. Ottawa, Ontario

4. Detroit, Michigan

5. Niagara Falls, Ontario

6. The city of Sydney is located on *what* prominent subdivision of Nova Scotia?

7. Which Canadian city is a major telephone and telecommunications centre?
   A. Brandon, Manitoba
   B. Lethbridge, Alberta
   C. North Bay, Ontario
   D. Moncton, New Brunswick

8. Which capital city has a Royal Canadian Mint that is the only one producing circulation coinage for Canada?

9. Which capital city is located on the Avalon Peninsula?

10. Which city is located farther north – London, England, or St. John's, Newfoundland?

11. Which city is farther north – Edmonton, Alberta, or St. John's, Newfoundland?

12. Fort William and Port Arthur amalgamated in 1970 to form what city?

13. Name the city that is the country's largest producer of automobiles.

14. Where are the Royal Botanical Gardens?

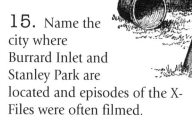

15. Name the city where Burrard Inlet and Stanley Park are located and episodes of the X-Files were often filmed.

16. Regina is to Saskatchewan as *what* is to New Brunswick?

17. Toronto is to Ontario as *what* is to Nova Scotia?

18. Does the CN Tower have a practical purpose?

19. Where is Canada's major West Coast ocean port?

20. Where is the major ocean port on Canada's East Coast?

21. What is the largest city between Toronto and Montreal on the Trans-Canada Highway?

22. C.M.A. stands for Census Metropolitan Area and includes only those cities with a population of 100,000 or more (city and suburbs included). How many C.M.A.'s are in Canada?

23. What percent of Canadians live in our C.M.A.'s?
- A. 30%
- B. 40%
- C. 50%
- D. 60%

24. What large Canadian C.M.A. city has the highest population of Aboriginal Canadians?

25. What American state has a population equal to or a little larger than the total population of Canada?

26. What is the total population of the Yukon, Nunavut, and the Northwest Territories?
- A. 50,000
- B. 100,000
- C. 150,000
- D. 200,000

27. Where was Western Canada's first jail?

28. How many people are currently living in Canadian prisons?
- A. 1 million
- B. 75,000
- C. 34,000
- D. 430,000

29. Which product is associated with Hamilton, Ontario?

30. What product is associated with the Chicoutimi-Lac St. Jean area of Quebec?

31. How quickly can you name this community?
Hint #1 – I am located at 46.5° north latitude and 81° west longitude.
Hint #2 – I was once struck by a meteorite.
Hint #3 – I am home to the Wolves.
Hint #4 – I have one of the world's largest deposits of chalcopyrite.
Hint #5 – Surveyors plotting Canada's first transcontinental railway accidentally discovered my hidden wealth.

COMMUNITIES

32. Where did Fredericton, New Brunswick, get its name?

33. Where did Charlottetown get its name?

34. What is the name of the most northern ice-free port in Canada?

35. What two Canadian cities are home to National Basketball Association teams?

36. Where is the Saddledome located?

37. Where is the Northlands Coliseum Place?

38. In the 1999-2000 season, the National Hockey League had six teams based in Canada. Which cities harbour these teams?

39. Long-time NHL fans speak of the 'Original Six' with great fondness. What cities did these six teams represent?

40. Hockey fans speak as if the pre-expansion 'Original Six' had existed since the beginning of time. How long were these six teams the only teams in the NHL?

41. If you were at the corner of Hollis and Sackville, what city would you be in?

42. What city is at the junction of the North and South Saskatchewan Rivers?

43. What city is at the junction of the Red and Assiniboine Rivers?

44. What city is at the junction of the St. Lawrence and Ottawa Rivers?

45. What city is at the junction of the Ottawa and Rideau Rivers?

46. What large Quebec City area waterfall is about 30 m higher than Niagara?

47. What city is at the confluence of the St. Charles River and the St. Lawrence River?

48. What city is built on the delta at the mouth of the Fraser River?

49. Does the St. Lawrence River have a delta?

50. Canada can lay claim to the world's longest street. Name it.

51. What city is closer to Saskatoon, Calgary or Edmonton?

52. In which city does the sun rise earlier, Regina or Calgary?

53. From which Western Canadian port do many cruises sail to Alaska?

54. What new Canadian territory was officially established on April 1, 1999?

55. What makes this Territory so unique in North America?

56. From where was this new Arctic territory formerly governed?

57. What settlement would you find at the mouth of the famous Coppermine River when fishing for Arctic char?

58. Changing place names can be very confusing for everyone – and most upsetting to cartographers. One Quebec community, for instance, is also commonly known by three other names. What is the official name of this community?

59. What are its other three names?

60. The Canadian National Exhibition is the largest annual exhibition in the world. Where is it held?

61. Where is the "Greatest Outdoor Show on Earth"?

62. Canada has hosted two very successful World Fairs. Where and when were they held?

63. What is the name of the Canadian organization that has been re-inventing the circus throughout the world?

64. *Attractions Canada* promotes Canadian tourism, annually selecting this country's top attractions. In 1999, winner for best national or international attraction was the Kings Landing Historical Settlement. Where is Kings Landing?

65. Ironically, *Attractions Canada's* choice for "best new attraction" featured rocks that are millions of years old. These rocks are found on the floor of a bay where they are completely exposed at low tide, thus allowing people to walk around them. What are these rocks called?

66. What's the name of the bay in which these rocks are found?

67. Also in 1999 – a good year for Atlantic tourist attractions – the Maritime Museum of the Atlantic was chosen best indoor tourist site in Canada. Where is this museum?

68. What famous sailing ship is often seen tied up at the wharf behind the Maritime Museum of the Atlantic?

69. What was judged to be the best outdoor site by *Attractions Canada* in 1999?

70. Where was Canada's first zoo founded?

71. In an Internet competition held in 1999, voters chose their overall favourite Canadian attraction. What attraction won?

72. It's said that people vote with their feet. How many millions of people visited Banff in 1998?
- A. 4.7
- B. 1.7
- C. 8.7
- D. 12.7

73. What Prairie city is home to a festival called "Islendingagurinn?"

74. In what city is the Peace Tower?

75. Which city hosted a summer Olympic Games?

76. Which city has hosted a winter Olympic Games?

77. What Canadian city hosted its second Pan-American games in 1999?

78. What Canadian cities have hosted the Commonwealth Games, or, as they were known earlier, the British Empire Games?

79. The Space Needle is to Seattle as the Skylon Tower is to *where?*

80. Which Eastern metropolis has a 233 m hill surmounted by an illuminated cross?

81. Which Eastern city's port is dominated by Signal Hill?

82. The Golden Boy sits atop the provincial legislature of which Prairie city?

83. Which Western city's skyline is dominated by The Empress?

84. Which Eastern city is dominated by the Chateau Frontenac?

85. What city is home to what used to be called Husky Tower?

86. Which Prairie capital city's Legislative building is located beside Wascana Lake?

87. Which Prairie city of over 200,000 citizens is bisected by the South Saskatchewan River?

88. Which Maritime city is home to the Tidal Bore?

89. Which Maritime coastal capital city hosted the Confederation Conference of 1864?

90. What was the last province to join Confederation?

91. What province would be missing on a map of Canada made in 1900 – B.C. or Saskatchewan?

92. Which province was first to join Canada – P.E.I. or Manitoba?

93. What four Canadian cities at one time or another housed Canada's capital?

94. Provinces are to Canada as *what* is to Switzerland?
   A. Cantons
   B. Bailiwicks
   C. Parishes
   D. Seigneuries

95. Two land systems were used when Lower Canada was first settled by French and British immigrants. The British used the Township System. What was the French land system called?

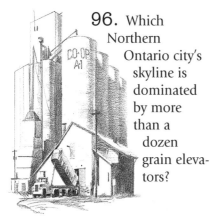

96. Which Northern Ontario city's skyline is dominated by more than a dozen grain elevators?

97. Which Northern Ontario city's skyline is dominated by the world's tallest smokestack?

98. What is Canada's largest grain port?

99. Stelco and Dofasco mills dominate part of what Ontario city's skyline?

100. Which Quebec pulp and paper city has a mis-numbered name?

101. Where is Percé Rock?

102. Where is Citadel Hill?

103. Where are the Reversing Falls?

104. What common Canadian place name means "shining waters"?

105. It's somewhat unclear where the word Canada originated. It's generally thought to have come from the Iroquoian term, "kanata." What does "kanata" mean?

106. Where is the "Golden Horseshoe"?

107. What city has the nickname, "Sandstone City"?

108. Where and what is the "Big Nickel"?

109. From what material is the "Big Nickel" constructed?

110. What is a "herring-choker"?

111. Who are "Bluenosers"?

112. In what lake does the monster Ogopogo supposedly live?

113. Who is Memphri?

114. Many Quebecers claim that they have sighted Loch Ness-like monsters in local lakes. From what part of Quebec do most of these claims stem?

115. What large fish has precipitated many of these Quebec sightings?

116. What is the Newfoundland peninsula named after King Arthur's mythical kingdom?

117. Where does the *Sasquatch* live?

118. What is the republic of Madawaska, and what is its honorary capital?

119. Where are the Frog Follies held?

120. Which town on the north shore of Lake Superior is home to a huge iron replica of a Canada Goose?

121. Where is the world's largest model lobster?

122. Where is the world's largest lobster trap?

123. Where is the world's largest replica of that famous freshwater game fish known as the Muskie?

124. On the same topic of huge fish, where is the world's largest trout reproduction – and the rod needed to land it?

125. Where is the world's largest salmon statue?

126. Still on the subject of fish, where is the world's largest channel catfish?

127. Where is the world's largest imitation moose?

128. Onanole, Manitoba, is home to a huge reproduction of one of the moose's relatives. What is it?

129. What town is home to a giant Tyrannosaurus Rex?

130. In the insect category, what town is home to the largest replica mosquito?

131. What huge imitation insect calls Wilke, Saskatchewan, home?

132. How about Falher, Alberta?

133. In the plant category, we have a flaming red lily replica located in which Saskatchewan community?

134. Where in Saskatchewan is the world's largest imitation pea plant?

135. Saskatchewan also boasts the world's largest imitation chokecherry. Where can it be found?

136. How about the world's largest model mushroom?

137. A Manitoba community claims to have the world's largest model of a sharp-tailed grouse, a category which probably has a short list of entrants. What's the name of the town?

138. A claim to the world's largest model whooping crane is led by what Saskatchewan town?

139. Where in Alberta is the world's largest man-made duck?

140. Where in Alberta is the largest great-blue heron?

141. How about the largest swan?

142. In the never-popular serpent category, which community is home to the world's largest mock garter snake?

143. What town in Saskatchewan boasts a synthetic wooly mammoth?

144. How about the world's largest gopher?

145. What large reproduction is found in Porcupine Plain, Saskatchewan?

146. In the rare animal parts category, in what town is the largest model of a rack of white-tailed deer antlers?

147. In the same category, where is the largest bunnock?

148. Switching categories, where is the world's largest oil can?

149. How about the largest cream can model?

150. Where in Saskatchewan is the world's largest coffee pot?

151. How about the world's largest tomahawk?

152. How about the world's largest *pysanka*, a Ukrainian-style decorated Easter egg?

153. Where is the largest pierogi?

154. How about the world's largest piggy bank?

155. And the largest roses?

156. What about the world's largest smoking pipe?

157. Where is the world's largest softball?

158. The world's largest teepee?

159. In which community is there a statue of a milk cow named Springbank Snow Countess?

160. And where is the world's largest model of Star Trek's *Starship Enterprise*?

161. Speaking of the *Enterprise*, Captain Kirk (William Shatner) was born in Canada. His alma mater even named a building in his honour. Name this university.

162. McMaster University is to Hamilton as Queen's University is to *where*?

163. In what city is the oldest Canadian university?

164. Where is Canada's newest degree-granting institution?

165. How many degree granting institutions are located in Halifax?

166. In which university did academic and television personality, David Suzuki, start his teaching career?

167. Only four educational facilities in Canada offer training in veterinary medicine. Where did the approximately 6,000 Canadian veterinarians receive their training?

168. Vancouver's Arthur Erickson is one of Canada's most internationally renowned architects. He is credited with designing all or part of at least three university campuses in Canada. Which three campuses have Erickson buildings?

169. Moses Coady began the Antigonish Movement, an economic self-help program for Maritime farmers and fishermen. The movement spread through much of the Third World. At which university did Father Coady teach?

170. The University of P.E.I. is to Charlottetown as Lakehead University is to where?

COMMUNITIES

171. University of Western Ontario is to London, Ontario, as Mt. Allison University is to where?

172. Memorial University is to St. John's as the University of New Brunswick is to where?

173. Laurentian University is to Sudbury as Bishop's University is to where?

174. What is Canada's most southern university?

175. What percent of Canada's total population are currently university students, part- or full-time?

176. What percent of the Canadian labour force has obtained a university degree?

177. Where was Canada's first English-language university established?

178. Which university was founded first, the University of Regina or the University of Saskatchewan?

179. Which university was founded first, the University of Alberta or the University of Calgary?

180. York University was founded in Toronto in 1959 with impetus from the North Toronto YMCA. It was originally affiliated with which one of the following Ontario universities:
A. University of Western Ontario
B. University of Waterloo
C. Trent University
D. University of Toronto

181. What other large Canadian urban university was founded with the help of the YMCA to promote adult education?

182. In 1968, York University became affiliated with what renowned Toronto law school?

183. Which Ontario community college has a world-recognized computer animation program whose graduates are snapped up by film animation production companies?

184. What weekly Canadian magazine publishes an annual ranking of Canadian universities and where is it published?

185. Where did the fourth International Geography Olympiad take place?

186. Where is "The Great Canadian Geography National Challenge" held each year?

187. What was Wiarton Willie?

188. How did Willie foretell the weather?

189. Name Willie's American equivalent and the originator of Groundhog Day.

190. What traumatic event took place at the Canadian Groundhog Day celebrations in 1999?

191. The *Titanic* collided with an iceberg and sank in international waters about 150 km east of the Grand Banks off Newfoundland. Many of the recovered bodies were buried in Canada. In what Canadian city are the most victims buried?

192. Who was the wealthiest and perhaps best known of the 250 Canadians who were on the *Titanic*?

193. What was the site of the world's largest man-made explosion until the atomic bombs of 1945?

194. What exploded?

195. How many people died?

196. Canada's worst West Coast marine disaster occurred in Alaskan waters. What ship was involved?

197. In 1949, the *Noronic* caught fire. What was it?

198. One of the most unfortunate Canadian marine disasters occurred in 1982 and didn't even involve a ship. What was the accident?

199. A fatal 1963 air crash killed 118 aboard Trans-Canada Airline Flight 831. Where did the plane depart from and where was it heading?

200. What was the cause of the crash?

201. Another DC-8 crashed in 1970 again after leaving Montreal for Toronto. Where did this aircraft crash?

202. What caused the crash?

203. What emerging city was destroyed by fire in 1886?

204. Perhaps the most serious single building fire occurred during a live radio broadcast from the Knights of Columbus Hotel in 1942. In what city was the fire?

COMMUNITIES

205. Where was the large theatre fire in 1927 that caused the death of 77 people, mostly children between the ages of 8 and 12?

Communities

Answers

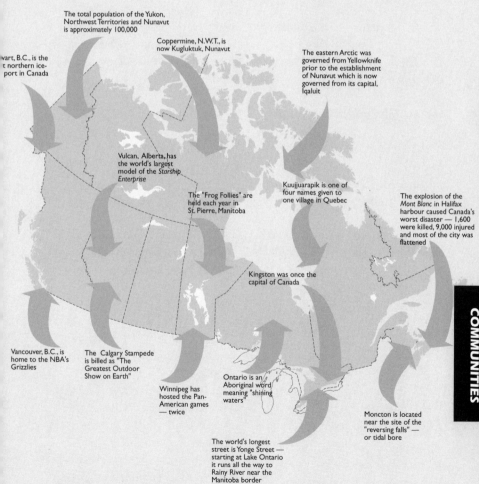

The total population of the Yukon, Northwest Territories and Nunavut is approximately 100,000

Coppermine, N.W.T., is now Kugluktuk, Nunavut

The eastern Arctic was governed from Yellowknife prior to the establishment of Nunavut which is now governed from its capital, Iqaluit

...vart, B.C., is the ...t northern ice-port in Canada

Vulcan, Alberta, has the world's largest model of the *Starship Enterprise*

The "Frog Follies" are held each year in St. Pierre, Manitoba

Kuujjuarapik is one of four names given to one village in Quebec

The explosion of the *Mont Blanc* in Halifax harbour caused Canada's worst disaster — 1,600 were killed, 9,000 injured and most of the city was flattened

Kingston was once the capital of Canada

Vancouver, B.C., is home to the NBA's Grizzlies

The Calgary Stampede is billed as "The Greatest Outdoor Show on Earth"

Winnipeg has hosted the Pan-American games — twice

Ontario is an Aboriginal word meaning "shining waters"

The world's longest street is Yonge Street — starting at Lake Ontario it runs all the way to Rainy River near the Manitoba border

Moncton is located near the site of the "reversing falls" — or tidal bore

COMMUNITIES

1. Kitchener.

2. Dartmouth.

3. Hull, Quebec.

4. Windsor, Ontario.

5. Niagara Falls, New York.

6. Cape Breton Island.

7. D. Moncton, New Brunswick

8. The Royal Canadian Mint in Winnipeg.

9. St. John's, Newfoundland.

10. London, England, located at 51.5° north latitude, is farther north, while St. John's is at 47.5° north latitude.

11. Edmonton is farther north. Edmonton is between 53° and 54° north latitude while St. John's is between 47° and 48° north latitude.

12. Thunder Bay.

13. Windsor, Ontario.

14. Hamilton, Ontario – the RBG is home to more species of lilac than any other botanical garden in the world.

15. Vancouver, B.C.

16. There are two answers. If you answered Fredericton, because both cities are provincial capitals, you are correct. If you answered St. John, because Regina is the largest city in Saskatchewan, then you are also correct.

17. Halifax (capital and largest city).

18. Aside from attracting tourists, the world's tallest free-standing structure serves as a base for communications equipment.

19. Vancouver.

20. Halifax.

21. Kingston, which has a population of 142,000.

22. Statistics Canada figures indicate that there are 25 C.M.A.'s in Canada.

23. C. 50%

24. Winnipeg. Its Aboriginal population totals over 35,000 people or 5.4% of the total population. Edmonton is home to the second highest population with a total of 29,425 or 3.5% of the total population. Regina and Saskatoon share the highest percent of total population figures with 5.7% each.

25. *According to the U.S. Census Bureau, the State of California has a population of over 32 million people, while Statistics Canada indicates that Canada's population is just under 31 million people. Two other states with large populations are Texas, with almost 20 million people, and New York, with a population of just over 18 million people.*

26. *B. 100,000*

27. *Emerson, Manitoba – it was erected in 1879.*

28. *C. 34,000*
Each prisoner costs Canadian tax-payers about $115 per day.

29. *Steel.*

30. *Aluminum.*

31. *Sudbury, Ontario. Sudbury is one of the most mineralized places on earth. Its richness is probably the result of a meteor which concentrated the minerals already in the earth into an oval shape marking the edge of impact. Chalcopyrite, a mineral which contains copper, is one of many minerals mined here. Sudbury is also home to the Wolves of the Ontario Major Junior Hockey League.*

32. *Frederick's Town was founded in 1785 by United Empire Loyalists as a "haven for the King's friends." It was named after Prince Frederick, second son of George III.*

33. *The community was origi-nally called Port-la-Joie but when ceded to Britain was renamed after Charlotte Sophia, queen of George III.*

34. *It is Stewart in B.C., right across a narrow body of water from the Alaska/U.S.A. boundary.*

35. *Vancouver, home of the Grizzlies, and Toronto, home of the Raptors.*

36. *Calgary, home of the NHL's Flames.*

37. *Edmonton, home of the NHL's Oilers.*

38. *Vancouver Canucks*
Edmonton Oilers
Calgary Flames
Toronto Maple Leafs
Ottawa Senators
Montreal Canadiens

39. *Toronto and Montreal in Canada, and Boston, New York, Detroit, and Chicago in the United States.*

COMMUNITIES

40. For twenty-eight seasons, from 1940 until 1967. Today, there are over two dozen teams with players from over a dozen nations.

41. Halifax.

42. Prince Albert, Saskatchewan.

43. Winnipeg.

44. Montreal.

45. Ottawa.

46. Montmorency Falls.

47. Quebec.

48. Vancouver.

49. Yes and no. It does not have a conventional river deposit at its mouth, which is the basic requirement of every delta. Rather, its mouth is a huge, wide, and deep estuary. Near Trois Rivières at the upstream west end of Lac St. Pierre, however, are river deposits which are typically deltaic.

50. Yonge Street in Ontario. Originally begun in 1795 to link York (Toronto) to Lake Simcoe to the north, the world's longest street now runs from Toronto to Rainy River, Ontario, near Lake of the Woods.

51. Edmonton is 531 km from Saskatoon and Calgary is 626 km from Saskatoon.

52. Since the sun appears first in the city which is further east, Regina, which is almost 10° east of Calgary, receives the day's first sun.

53. Vancouver.

54. Nunavut Territory.

55. It is governed by a predominantly aboriginally-elected government.

56. This territory was part of the Northwest Territories, which was governed by the Federal Government through the capital in Yellowknife.

57. The community once called Coppermine is now Kugluktuk, Nunavut. Many place names have reverted to original First Nations versions.

58. Kuujjuarapik is the Inuit name for this town on the Hudson Bay.

59. The Cree call it Whapmagoostui, English-speakers call it Great Whale, and to French-speakers it is known as Grande Baleine.

60. *Toronto is the host of this annual event which first opened in 1846 as an agricultural fair. It now offers a diverse range of rides, amusements, food, concerts, and attracts millions of people each summer. The exhibition is now known simply as the CNE, or the Ex.*

61. *This self-promoting title describes the Calgary Stampede. The millions who visit each year are not disappointed. This huge fair and rodeo starts in early July of each year and runs for ten days.*

62. *Montreal hosted Expo 67 in 1967 to celebrate Canada's centennial. This Class One exhibition drew over 50 million visitors in the 6 months it was open. Vancouver played host to Expo 86 in 1986. This Fair ran for approximately 6 months, drawing over 20 million visitors.*

63. *Cirque du Soleil – although performers hail from around the world, the organization's homebase is Montreal.*

64. *In New Brunswick, just 30 km north of Fredericton on the Trans-Canada Highway. This re-creation of a Loyalist settlement beat out the Calgary Stampede and the Montreal International Jazz Festival.*

65. *The Hopewell Rocks in New Brunswick. Generally known as flower pots, these rocks are mush-room-shaped columns of sand-stone that have been shaped by the tides. The rocks themselves aren't new, of course, but the 1998 facilities for viewing them are.*

66. *The Bay of Fundy, located between New Brunswick and Nova Scotia. The rocks are located a short drive from Moncton on the New Brunswick side.*

67. *At Harbourside, on Lower Water Street, Halifax.*

68. *The Bluenose II, which is a 1963 replica of the original Bluenose schooner of 1921. Both ships were built at the same shipyard in Lunenburg, Nova Scotia.*

69. *The world famous Calgary Zoo, which has extensive gardens containing over 10,000 plants, a Canadian-wilderness exhibit, a dinosaur park complete with 20 life-size replicas, and a first-class list of rare and exotic animals.*

70. *North America's first zoo opened in Halifax in 1847.*

71. *The winner was Banff National Park in Alberta.*

72. *A. 4.7*

COMMUNITIES

73. Gimli, Manitoba – the Icelandic community celebrate this festival with a parade, music, and food.

74. Ottawa.

75. Montreal in 1976.

76. Calgary in 1988.

77. Winnipeg is the only Canadian city to have hosted these games. They were first held there in 1967.

78.
Hamilton – 1930
Vancouver – 1954
Edmonton – 1978
Victoria – 1994

79. Niagara Falls, Ontario. The tower is 160 m above ground and 244 m above the base of the falls.

80. The Cross of Christ is located on Mount Royal, Montreal.

81. St. John's, Newfoundland.

82. Winnipeg. This 4 m tall statue weighs 5 tonnes, and is covered in 24-karat gold.

83. Victoria. This majestic Canadian Pacific hotel was completed in 1908.

84. This CP hotel was completed in 1893 and looks across the St. Lawrence from Quebec City.

85. Calgary. The tower is now called Calgary Tower.

86. Regina.

87. Saskatoon.

88. Moncton – the Tidal Bore occurs on the Petitcodiac River in New Brunswick. The Bore is a small wall of water that heads upstream when the volume of water in the river is lower than the volume of the Bay of Fundy at high tide. Essentially, the water in the Bay overcomes the river flow and flow direction is reversed for a short time.

89. Charlottetown.

90. Newfoundland, in 1949.

91. Saskatchewan, which became a province in 1905. B.C., on the other hand, joined Canada in 1871.

92. Manitoba, which entered in 1870. P.E.I. joined Confederation in 1873.

93. Kingston, Toronto, Montreal, and Quebec.

94. A. Cantons

95. *Names variously attached to it are: Rang System, Long Lot System and Seigneurial System.*

96. *Thunder Bay.*

97. *Inco's superstack in Sudbury is 381 m high. (The CN Tower in Toronto is 555 m tall.)*

98. *Vancouver, with over 45% of all Canadian grain exports.*

99. *Hamilton.*

100. *Trois Rivières, at the confluence of only two rivers, the St. Lawrence and the St. Maurice. An island at the mouth of the St. Maurice gives the false impression that two rivers flow into the St. Lawrence, hence the mis-numbered name.*

101. *The Gaspé Peninsula, Quebec.*

102. *The Citadel is a large fortress built by the British in the nineteenth century to help defend Halifax. The fort stands on a large hill of glacial debris called a drumlin.*

103. *Saint John, New Brunswick.*

104. *The Aboriginal name for "shining waters" is Ontario.*

105. *Village.*

106. *This region along the west end of Lake Ontario is marked by a string of cities running from Oshawa to Niagara Falls. The region includes Toronto and Hamilton, and is noted for its dense population and high manufacturing concentration.*

107. *Calgary adopted a by-law in 1886 mandating that new downtown buildings be made of sandstone. This law was the result of a fire that destroyed the wooden buildings which formerly made up the downtown area. Calgary's downtown now boasts the normal glass and steel construction but older buildings still show evidence that sandstone was once the most common building material.*

108. *The Big Nickel is a massive replica of a Canadian 5-cent piece measuring 7 m in diameter. It sits on top of a popular demonstration mine depicting the mining process in Sudbury, on a hillside near the Trans-Canada Highway.*

109. *"The Big Nickel" is constructed from non-rusting stainless steel (made by adding nickel to molten steel).*

COMMUNITIES

110. *It is a slang term for a New Brunswicker. The term stems, perhaps, from the large number of herring fishermen who live along the Bay of Fundy.*

111. *Nova Scotians, usually those from the South Shore, or the Atlantic coast west of Halifax. The origin of the name is unclear. Some suggest it refers to the noses of fishermen turned blue from the cold; others suggest that it stems from the Scottish Presbyterians who were referred to as "true blue" in the seventeenth century. Then again, the blue may be from the dye in mittens used to wipe runny noses when out fishing.*

112. *Lake Okanagan, B.C.*

113. *Perhaps "what" is the better word. Memphri is the "sea-monster" resident of Lake Memphremagog, the large border lake between Quebec and Vermont.*

114. *Most Quebec "sightings" have occurred in the Cantons de l'Est or the Eastern Townships, especially to the east near the isolated Maine border. There are many deep lakes in this area, ideal for large, aquatic monsters.*

115. *Over the years, large sturgeon weighing more than 225 kg were commonly caught in*

Quebec waters. Their skin, with bony plate-like dorsal fins, lead many people to assume that these large, prehistoric-looking fish are, in fact, monsters.

116. *Avalon Peninsula.*

117. *The legendary Big Foot stalks the Pacific Northwest. Infrequent sightings lend some credence to the existence of this large, hairy, ape-like creature.*

118. *Madawaska is a mythical kingdom on the Maine/New Brunswick border. Its most famous citizen is Paul Bunyan. Edmundston, New Brunswick, is Madawaska's honorary capital.*

119. *The Frog Follies, featuring the Canadian frog-jumping contest among other attractions, are held every July in St. Pierre, Manitoba.*

120. *Wawa, Ontario.*

121. *Shediac, New Brunswick.*

122. *Cheticamp, on Cape Breton Island.*

123. *Kenora, Ontario.*

124. *The trout is in Kamloops, B.C., but one would have to go to another B.C. community, Houston, to find the immense fly-*

fishing rod needed to land the Kamloops trout.

125. *Campbellton, New Brunswick.*

126. *Selkirk, Manitoba.*

127. *Mac the moose "lives" in a fitting location, Moose Jaw, Saskatchewan. He is approximately 20 times the weight of an actual bull moose.*

128. *A seven-point fibreglass bull elk.*

129. *Dini, all 10 plus metres of him or her, is made of concrete and is on display in Drumheller, Alberta.*

130. *Komarno, Manitoba, although many Manitoba residents claim to have been bitten by larger, real ones.*

131. *The grasshopper.*

132. *The honeybee.*

133. *Parkside.*

134. *St. Isadore du Bellevue.*

135. *Lancer.*

136. *Vilna, Alberta.*

137. *Ashern.*

138. *Govan.*

139. *Andrew.*

140. *Barrhead.*

141. *Grande Prairie.*

142. *Inwood, Manitoba.*

143. *Kyle.*

144. *Eston, Saskatchewan.*

145. *What else, but a porcupine.*

146. *Gorlitz, Saskatchewan.*

147. *What do you mean, what's a bunnock? It's a horse's ankle bone and the largest is found in Macklin, Saskatchewan.*

148. *Rocanville, Saskatchewan.*

149. *Marker, Alberta.*

150. *Davidson.*

151. *Fittingly, sort of, the giant tomahawk is located in Cut Knife, Saskatchewan.*

152. *Vegreville, Alberta.*

153. *Glendon, Alberta.*

154. *Coleman, Alberta.*

155. *Roseisle, Manitoba, of course.*

156. *St. Claude, Manitoba.*

157. *Chauvin, Alberta.*

158. *Medicine Hat, Alberta.*

159. *Woodstock, Ontario. The countess was the first cow of any breed to set six butterfat world records. She lived in the 1920's.*

160. *Vulcan, Alberta. Mr. Spock, a Vulcan, would approve.*

161. *McGill University in Montreal.*

162. *Kingston.*

163. *Quebec City, where Le Séminaire de Québec was founded by the Catholic Church through Monseigneur François de Laval as a college of fine arts and theology in 1663.*

164. *Prince George, B.C., is home to the recently founded University of Northern B.C.*

165. *Five: Dalhousie University; St. Mary's University; Mount St. Vincent University; Nova Scotia College of Art & Design; University of King's College. Technical University of Nova Scotia has recently become part of Dalhousie.*

166. *Genetics students at the University of B.C. in Vancouver benefited from Suzuki's research and teaching.*

167. *Western College of Veterinary Medicine, Saskatoon; Ontario Veterinary College, Guelph; Atlantic Veterinary College, Charlottetown; L'Université de Montréal, St. Hyacinthe campus, Quebec.*

168.
a) *Simon Fraser University in Burnaby, B.C.*
b) *University of Lethbridge, Alberta.*
c) *University of B.C., Vancouver.*

169. *Father Moses Coady was the first director of the Extension Department of St. Francis Xavier University in Antigonish, Nova Scotia. St. F.X. continues the work of helping people help themselves through the Coady International Institute.*

170. *Thunder Bay, Ontario.*

171. *Sackville, New Brunswick.*

172. *Fredericton, New Brunswick.*

173. *Lennoxville, Quebec.*

COMMUNITIES

174. *University of Windsor, Windsor, Ontario.*

175. *According to Statistics Canada, close to 2% of Canadians presently attend university.*

176. *Just under 14% of the work force has obtained a Bachelor's degree or higher.*

177. *This honour goes to the University of New Brunswick, which was founded in 1785 as the Provincial Academy of Arts and Science. It received its charter in 1800.*

178. *The University of Saskatchewan in Saskatoon was founded in 1907. It evolved from Emmanuel College in St. Albert, which was founded in 1879. Regina College, on the other hand, was originally a junior college affiliated with the University of Saskatchewan. It was given degree-granting status in 1959 as a campus of the University of Saskatchewan. In 1974, it became the autonomous University of Regina.*

179. *The University of Alberta was founded at the first sitting of the Alberta Legislative Assembly in 1906. The University of Calgary was founded in 1966, but began as a branch of the*

Faculty of Education of the University of Alberta in 1946.

180. *D. University of Toronto*

181. *Sir George Williams University in Montreal, which is now part of Concordia University. It was founded by the "Y" in 1873.*

182. *Osgoode Hall Law School, founded in 1862.*

183. *Sheridan College. Its two campuses are located in Brampton and Oakville, both west of Toronto.*

184. *Maclean's magazine is published in Toronto. Its French-language counterpart is L'Actualité, a spin-off from Le Maclean of the 1970's, published in Montreal.*

185. *Toronto in August, 1999 – Eleven countries competed. The United States won gold, Canada won silver and Russia won bronze. It was sponsored by the National Geographic Society and the Royal Canadian Geographical Society.*

186. *Ottawa, over the Victoria Day weekend.*

187. *A groundhog from Wiarton, Ontario, and an indicator of how much more winter weather was due after February 2, Groundhog Day.*

188. *If Willie came out of his burrow and saw his shadow, we were in for six more weeks of winter weather, which in Canada isn't so bad. Spring was coming if he didn't see his shadow. If he went back into his hole, more bad weather was coming. If he remained above ground, spring was near. Willie's handlers claimed that his predictive abilities were virtually 100% accurate. An impartial jury might decide otherwise.*

189. *Punxsutawney Phil of Punxsutawney, Pennsylvania, who, along with his offspring, have been in the weather prediction business for over a century.*

190. *It was announced that Willie had passed away a few days earlier and Willie's albino corpse was displayed as proof. As it turned out, the corpse was not Willie but a nameless stand-in who smelled better than the defunct herbivore. Evidently, Willie's stand-in had been refrigerated. Poor Willie had not.*

191. *150 victims are buried in Halifax.*

192. *Harry Markland Molson, banker and owner of Molson breweries. At the time of his death, Molson was mayor of Dorval, a suburb of Montreal.*

193. *The Halifax Explosion occurred on December 6, 1917.*

194. *The fully loaded French munitions ship, Mont Blanc, was struck by the Norwegian freighter, Imo, and caught fire in the Halifax harbour.*

195. *More than 1,600 people were killed and over 9,000 injured. Most of the city of 50,000 inhabitants was flattened.*

196. *On October 24th, 1918, the Canadian Pacific ship Princess Sophia ran aground on Vanderbilt Reef in Lynn Canal during a fierce blizzard while sailing from Skagway, Alaska, to Vancouver. Rescue ships were unable to reach the boat and, although within sight of land, all 343 aboard drowned.*

197. *The Noronic was a Great Lakes cruise ship. It caught fire at dock in Toronto, killing 118 passengers in their sleep.*

198. *It was the Valentine's Day sinking of the oil rig Ocean Ranger during a fierce storm. Eighty-four crew members lost* their lives, as the rig sank off the east coast of Newfoundland in the Grand Banks.

199. Flight 831 departed from Dorval Airport near Montreal destined for Toronto. It crashed in Ste. Thérèse, just a few km north of Dorval Airport.

200. Nobody knows for certain because all aboard were killed and the aircraft was not carrying a black box (flight data recorder). After 1965, black boxes, along with Cockpit Voice Recorders, became mandatory on all planes.

201. In a farmer's field, just outside the Toronto airport. The plane made its approach without problem, but suddenly dropped 15 m and hit the runway. The pilot pulled away and flew a short distance after initial contact but three explosions caused a wing to fall off. One hundred and nine people perished.

202. A newly installed "black box" recorded the cockpit conversation and provided evidence that the co-pilot wrongly deployed wing spoilers. Similar crashes caused the aircraft manufacturer to change the method of wing spoiler adjustment.

203. Vancouver – a small fire started on the edge of town and quickly engulfed the whole town. The fire wasn't extinguished; it burned itself out. Vancouverites started re-building the same day.

204. St. John's, Newfoundland. Ninety-nine people were killed, mostly Canadian and American servicemen and their girlfriends.

205. The Laurier Palace Theatre was in Montreal.

Canadian Geographic Quiz Book

People

Art
Literature
Film & Photography
Famous People
Forward Thinkers
Music
Sports
History & Population
Explorers

1. Where and what is the Glenbow?

2. One of Canada's most significant paintings is *The Fathers of Confederation*. The finished original was lost in the 1916 fire that destroyed the Parliament Buildings. The artist's preparatory sketches are all that remain today. Where do they hang?

3. Who painted this group portrait and where was he born?

4. What Ontario art museum is dedicated to the works of the Group of Seven and where is it located?

5. The Group of Seven was famous for depicting what in their paintings?

6. Who were the original members of the Group of Seven?

7. What famous painter of this period is often mistakenly included in the Group of Seven?

8. How and where did Tom Thomson die?

9. The artists of *what* Quebec town are famous for their traditional wood carvings?

10. Emily Carr, recognized as one of Canada's leading artists, led a very varied career. Just as she started to receive recognition in her home province she was abandoned by her family. Carr did not paint for the following twenty years of her life.  She kept a boarding house and made pottery for tourists. Eventually because of Lawren Harris's encouragement she resumed painting, but it was not until Carr entered her seventies that she received her just due as a painter. Where was Emily Carr born and where did she live most of her life?

11. What town on the west coast of Vancouver Island is home to a burgeoning artist colony?

12. Toronto-born Alex Colville gained fame as a World War II artist. Where did he teach art on his return from Europe?

13. Christopher Pratt calls what part of Canada home?

14. What Canadian artist painted *At the Crease* and is renowned for his realistic sports scenes?

15. Who is Canada's most famous nature painter?

16. William Kurelek had a tumultuous career cut short when he was fifty years of age. He was a prolific painter and is credited with finishing over 2,000 paintings. Kurelek also wrote and illustrated a number of stories for young Canadians. What part of Canada was most featured in his work?

17. Paul Kane was a well-known artist born in Ireland in 1810. He was brought up in Toronto and gained fame for his everyday depictions of Aboriginal Canadians. What region of Canada did Kane favour in his paintings?

18. Cornelius Krieghoff was a contemporary of Kane's. Both men painted highly detailed scenes of contemporary life in oils. What area of Canada did Krieghoff feature most in his work?

19. Who is Ozias Leduc and where did he live?

20. Renowned Métis architect, Douglas Cardinal, designed the $93 million Canadian Museum of Civilization in Hull, Quebec. His magnum opus, however, is commonly thought to be St. Mary's Church, located *where*?

21. Who is Kenojuak and where is she from?

22. *Raven and the First Humans* is a masterpiece created by sculptor Bill Reid. Where is it located?

23. In what Canadian city can you find a large collection of British sculptor Henry Moore's work?

24. *Maria Chapdelaine* is a famous Canadian novel about the difficulties of nineteenth-century pioneer life and one's attachment to native soil. Who wrote the book and where was the author born?

25. Where was the author of *Maria Chapdelaine* killed?

26. Lucy Maud Montgomery is certainly P.E.I.'s most successful novelist. Her *Anne of Green Gables* manuscript, however, was rejected by every Canadian publishing house she submitted it to. Where was it published after three years of rejections?

27. In what province did Lucy Maud Montgomery write the majority of her beloved "Anne" novels?

28. Most Canadians are familiar with John McCrae's poem, "In Flanders Fields." About what war did he write?

29. Where was McCrae born and educated, and what was his occupation while in the Army?

30. Dennis Lee is author of *Alligator Pie*, a popular book of children's poetry, and *Civil Elegies and Other Poems*, which received the Governor General's Award for Poetry in 1972. Where is he from?

31. "David" is one of Earle Birney's best known poems. It tells the tale of two friends climbing and hiking in the mountains. Which mountains?

32. Like many poets and novelists, Earle Birney taught at his alma mater. Name this university.

33. Alice Munro often sets stories in Jubilee, a fictional town based on the small southwestern Ontario town in which she was born and raised. What's the name of this real town?

34. Where did Margaret Laurence set her first two novels, *The Tomorrow-Tamer* and *This Side Jordan*?

35. What fictional town did Margaret Laurence make famous in some of her novels?

36. It is hard to think of a more prolific writer of poetry than Irving Layton. He is credited with producing more than 50 volumes of English verse, despite the fact that he was not born in an English-speaking country. Where was he born?

37. Where did Layton teach?

PEOPLE

38. Leonard Cohen is a multi-talented artist. He has written poetry, songs, novels, is a noted stage performer, and has released more than a dozen albums. Where was Leonard Cohen born and raised?

39. Saul Bellow, noted American novelist, Pulitzer Prize winner and Nobel laureate, was born and raised in what Canadian city?

40. What Nova Scotian author and creator of Sam Slick coined such sayings as "stick in the mud," "jack of all trades and master of none," "six of one, half a dozen of another," and "upper crust?"

41. Most know that Hugh MacLennan, author of *Barometer Rising* and *Two Solitudes,* grew up in Halifax. Where was he born?

42. Where did MacLennan teach in Montreal?

43. Where did Farley Mowat grow up?

44. Where is Farley Mowat's book, *Lost in the Barrens,* set ?

45. What is the title of the Farley Mowat book which chronicles his visits to the Union of Soviet Socialist Republics, and what price did he pay for his praise of Soviet life?

46. Prolific chronicler, Pierre Berton, calls what part of Canada home?

47. In her novel, *Obasan,* Joy Kogawa wrote of her childhood experiences as a Japanese-Canadian forcibly removed to the B.C. interior. From where was she removed and where did she live for most of World War II?

48. Where was famous Prairies author W.O. Mitchell born?

49. W.P. Kinsella gained fame for writing two diverse groups of novels and collections of short stories. What are the two topics on which he focused?

50. What famous movie starring Kevin Costner is based on *Shoeless Joe?*

51. Actor and novelist, Gordon Pinsent, calls what small town home?

52. What is the setting of Gordon Pinsent's novels, *The Rowdyman* and *John and the Missus?*

53. Robertson Davies' *Deptford Trilogy* novels were set in what small Ontario town?

54. After attending high school at Upper Canada College in Toronto, Davies attended university in the town where his family lived. Name this town.

55. What imaginary town is the setting of Davies' trilogy comprised of *Tempest-Tost, Leaven of Malice,* and *A Mixture of Frailities?* (Hint: the name of the town is also the title of the trilogy.)

56. In what year was the Stratford Festival established?

57. What long-time Canadian novelist was also a successful actor and appeared at the first Shakespearian Festival in Stratford, Ontario?

58. Which Canadian novelist wrote, amongst other works, *The Many Coloured Coat* and *A Fine and Private Place?*

59. The above-mentioned author worked summers at the *Toronto Star* when he was a uni-versity student. There he met another reporter – an American novelist – who encouraged him to write fiction. Who was this American *Star* employee?

60. *Cabbagetown* is a novel about a working-class neigh-bourhood in Toronto. Who wrote it?

61. One of Canada's best-known authors lives in Toronto. This person has produced close to a dozen poetry collections, over half a dozen novels, children's books and several works of literary criticism. Who is this writer?

62. The novel *Life Before Man* is set where?

63. What Michael Ondaatje novel was turned into a screen-play which won an Academy Award for Best Picture?

64. Where was Michael Ondaatje born?

65. What city is the setting for Anne Carson's, *Autobiography of Red?*

66. Rudy Wiebe's, *The Temptation of Big Bear*, won a Govenor General's Award for Fiction in 1973. Wiebe is an emphatic exponent of Aboriginal Canadian issues although he is not an Aboriginal himself. What part of Canada is most featured in Wiebe's writing?

67. Guy Vanderhaeghe, noted for his writings set in Western Canada, won the 1996 Governor General's Award for Fiction for his novel, *The Englishman's Boy.* Where was Vanderhaeghe born?

68. Prairie novelist, James Sinclair Ross, left school early in life to work for the Royal Bank of Canada, writing around this "other" job. Ross became very familiar with the Prairies because all his bank postings, with the exception of one in Montreal, were in Prairie towns. Where was Ross born?

69. Characters in Ross's novels often surmount a typical Prairie hazard in order to triumph. What is this hazard?

70. What Yukon resident submitted his book of poems to the forerunner of Ryerson Press in 1907, expecting that he would have to pay for its publication personally?

71. What geographical feature was immortalized in the novel, *The Trail of '98?*

72. Frederick Philip Grove lived the last half of his life in Western Canada. A prolific poet and novelist, Grove died in 1948. Many years after his death, it was discovered that he had started his literary career as a poet, novelist and translator in another country. When he and his wife decided to emigrate, Grove staged his own death and started a new life in Canada. What was his country of birth?

73. Brian Moore, author of *Judith Hearne*, immigrated to Canada in the late forties from where?

74. Some people maintain great abilities in various disparate fields and therefore never cease to impress their contemporaries. Such a man was F.R. Scott who, like Morley Callaghan, was both a lawyer and an author. As a lawyer, Scott fought many interesting political battles with the governments of his day. Where did he practice law?

75. Where was F.R. Scott born?

76. Mazo de la Roche is best remembered for her internationally acclaimed *Jalna* stories about the Whiteoak family. These stories were released over a period of 33 years and were even adapted for television. What city did de la Roche call home?

77. Antonine Maillet's novels are rooted in the Acadian culture of her upbringing. Where was her childhood home?

78. Archie Belaney was an Englishman who married an Iroquois woman named Anahareo and travelled in North America and Europe under an assumed Aboriginal identity. Belaney's most popular book was *Pilgrims of the Wild*. What was his pseudonym?

79. Belaney worked and lived in what two national parks?

80. Anne Hébert wrote a novel published in 1970 about the murder of a young seigneur by his wife. The title is the name of the seigneury and the town in which the murder took place. What is it?

81. Tomson Highway is a Manitoba playwright noted for his depictions of reserve life. Two of his most famous plays, *Dry Lips Oughta Move to Kapuskasing* and *The Rez Sisters*, are set on the Wasaychigan Hill Indian Reserve. Where is this reserve?

82. Where was French-Canadian novelist Gabrielle Roy born?

83. E. J. (Ned) Pratt is one of Canada's best-known poets. Where was he born and raised?

84. What was *Tish*?

85. Who wrote a short story about a young Quebec boy who wanted a Canadiens hockey jersey but, because of a mix-up in the catalogue mailing, receives instead a Toronto Maple Leafs sweater?

86. Where was the first movie theatre in Canada established?

87. Philip Borsos' film, *The Grey Fox,* is set in what B.C. city?

88. Who played the role of Norman Bethune in the 1987 feature film, *Bethune*?

89. What was Bethune famous for?

90. Where was Norman Bethune born?

91. Al Waxman portrayed the "King" of what Toronto neighbourhood?

92. The television series *Due South* revolves around a Mountie fighting crime in what American city?

93. Where was the silent-movie comedian, Mack Sennett, from?

94. Winnie-the-Pooh is associated with what Canadian city?

95. Movie buffs know that three different Canadian-born actresses won consecutive Academy Awards for Best Actress in 1929, 1930, and 1931. Who were they and where were they born?

96. Who has won more Academy Awards than any other Canadian?

97. What is perhaps Canada's most famous family-run photographic studios?

98. Who is Canada's most famous portrait photographer and what are his geographical origins?

99. One of the signers of the *Declaration of Independence* was in charge of the American forces occupying Montreal in 1775-76, and founded the predecessor of a Canadian daily newspaper. Who was he?

100. Charles Dickens visited Canada in 1842 where he made the following observation about *what* town: "Full of life and motion, bustle, business and improvement. The streets were well paved and lighted with gas; the houses are large and good; the shops excellent"?

101. Dickens said this about what other town: "One half of it appears to be burnt down, and the other half not to be built up"?

102. When Samuel Clemens, a.k.a. Mark Twain, visited Canada in the nineteenth century, he had this to say about what city: "It is the first time that I am in a city where if I were to throw a rock in any direction, I could scarce avoid breaking a church window"?

103. Who was responsible for building the trans-continental railroad?

104. Wealthy New Brunswick industrialist, K.C. Irving, was reportedly worth more than $6 billion on his death which placed him among the top 20 wealthiest people of his day. He lived most of his last twenty years outside Canada in an effort to avoid taxes. Where was his tax haven?

105. Millionaire philanthropist Izaak Walton Killam was born in what city?

106. Alfred Fuller, founder of the Fuller Brush empire, hailed from what Eastern city?

107. Where did Alphonse Desjardins found the first credit union in 1900?

108. Conrad Black is a multi-millionaire owner of more than 500 newspapers worldwide. Where was his first newspaper based?

109. Samuel Cunard, son of the founder of the famous shipping line bearing his name, was born where?

110. Eaton's department store chain was a Canadian institution for more than a century. Its founder, Timothy Eaton, was not born in Canada. Where did he emigrate from?

111. Where was the first T. Eaton store located?

112. In 1944, immunologist Oswald Avery declared that DNA — not protein — was in fact the controlling molecule of life. Where was Avery born?

113. Charles Saunders developed what strain of wheat at the Dominion Experimental Farms in Ottawa?

114. At what university did Frederick Banting, C.H. Best et al. discover insulin, an effective therapy for diabetes mellitus?

115. Where in Canada was the nutritionally-balanced food for toddlers known as Pablum invented?

116. Harold Innis was an internationally recognized historian and political scientist best remembered for his economic staple thesis and classic 1930 history, *The Fur Trade in Canada*. Innis spent his entire academic career teaching at *what* university?

117. Who was Canada's first licenced female doctor?

118. Where did she receive her medical training?

119. Where did Alexander Graham Bell invent the telephone?

120. Alexander Graham Bell also developed the first powered aircraft in Canada. Where did it fly?

121. Another of Bell's many creations was the hydrofoil boat. Where did it set the 1919 world speed record?

122. Frederick Gisburne is credited with laying North America's first undersea telegraph cable in 1852. What parts of Canada did this cable link?

123. Guglielmo Marconi received the first trans-Atlantic wireless message on a hilltop in what Eastern capital?

124. American industrialist Cyrus Field is credited with being the moving force behind the first communications cable connecting Europe and North America. What two places did the cable connect?

125. The world's first radio broadcasts sent via radio waves were transmitted by Canadian Reginald Fessenden. The transmission was made from Massachusetts to United Fruit Company Ships on December 24, 1906. What part of Canada did Fessenden call home?

126. Where was the first electric radio developed?

127. The Lazer is a small single-person sailboat which has introduced thousands around the world to the pleasures of sailing. Where did Bruce Kirby, Hans Fogh, and Ian Bruce first sail and later produce thousands of the craft?

128. Where was the controversial CF-105 Avro Arrow jet fighter developed and built?

129. The rugged bush plane is a Canadian invention, and the dependability of the prototype became the standard by which all others were judged. Who developed the first bush plane and where?

130. The variable pitch propeller was considered a landmark invention. It allowed pilots to change the angle of propeller blades to compensate for varying atmospheric conditions. Where was this innovation conceived?

131. What are the Snowbirds, and where are they based?

132. Where was the first integrated school in North America founded?

133. Who founded it?

134. Henri Bourassa was an early Quebec nationalist, federal MP, and founder of *Le Devoir*, a daily newspaper that still reflects the initial views of its founder. Where is *Le Devoir* published?

135. René Lévesque, the modern-day father of Quebec's nationalist movement, was born in which province?

136. Paul-Émile Léger was born in Valleyfield, Quebec, and became the first Canadian Cardinal of the Roman Catholic Church in 1953. In 1967, he resigned his position as Cardinal, became a missionary, and worked with lepers in what area of the world?

137. The American ship, *Edmund Fitzgerald*, came to a tragic end on November 10, 1975, in *which* of the Great Lakes?

138. Which Great Lakes state has a memorial park dedicated to the *Edmund Fitzgerald*?

139. "The Wreck of the Edmund Fitzgerald" was written and recorded by whom?

140. What mountain range and river does Gordon Lightfoot pay homage to in "The Canadian Railway Trilogy"?

141. What song written and performed by *The Tragically Hip* refers to a line of longitude?

142. What city do *The Tragically Hip* call home?

143. Which famous actor/comedian and friend of the *Hip* also calls Kingston home?

144. Maureen Forester, the world famous contralto, is perhaps Canada's most celebrated opera singer. In 1986, she was named Chancellor of what university?

145. Who recorded Canada's first classical guitar music in 1975?

146. Popular entertainer, Susan Aglukark, lives in what city?

147. What Alberta-born singer/songwriter sang at the close of the Calgary Olympics?

148. Famous Quebec chansonnier, Gilles Vigneault, comes from what small fishing village on the St. Lawrence?

149. Where was Louis Quilico, the Canadian opera baritone, born?

150. Where did "supergroup" *The Guess Who* come from?

151. *The Guess Who*'s 1969 hit *American Woman* was the first Canadian single to do what in the United States?

152. What Western Canadian city is cited in the title of a *Guess Who* song?

153. Glenn Gould, born in 1932, began his relatively short musical career as a child prodigy but died at the early age of fifty. At age 15, Glenn Gould debuted with his hometown symphony orchestra. Which one?

154. What instrument did Glenn Gould play and in his idiosyncratic fashion "hum along to"?

155. Canada's most famous jazz musician is Oscar Peterson. In the early 1990's he was appointed Chancellor of what university?

156. Guy Lombardo and his *Royal Canadians* formed one of the most famous big bands of the 50's. From where did Guy Lombardo hail?

157. In addition to his skills as a bandleader, Guy Lombardo was associated with what sport?

158. Where were the two 1950's "guy groups" the *Diamonds* and the *Crew Cuts* from?

159. Who has won the most Juno awards and where was this singer born?

160. Bobby Gimby earned the nickname the "Pied Piper of Canada" for his official Canadian centennial theme hit song, *CA-NA-DA*. He also appeared on *Happy Gang* from 1948 to 1959. Where was Bobby Gimby from?

161. *Lighthouse*, the Canadian horn-driven band of the late 60's and mid 70's, made a poor career choice when they refused to play what famous summer of '69 concert?

162. In what Ontario town was the inventor of basketball, James Naismith, born?

163. The first Arctic Winter Games was held in 1970 in what city?

164. Terry Fox's cross-country run to support cancer research and awareness was tragically cut short. Where was his westward trek from St. John's to Victoria halted?

165. Where was famous nineteenth-century strongman Louis Cyr from?

166. Marilyn Bell was the first person to swim across which Great Lake?

167. Where was boxing great George Chuvalo born?

168. Who was the first Canadian to win an Olympic Gold and where was this medal won?

169. Canadian women made their debut in the first co-ed Olympics in 1928. However, the first Canadian woman to hold a world athletic title, speed skater Lela Brooks, won three world titles at the 1926 World Championships. Where was Lela Brooks from and where in Eastern Canada were the 1926 world speed skating championships held?

170. Most sports fans know that Jackie Robinson broke professional baseball's colour barrier. Where did he start his professional baseball career?

171. Where is Hall of Fame baseball pitcher Ferguson Jenkins from?

172. Ferguson Jenkins played most of his career in Chicago with the Cubs. What other major league teams did he play for?

173. Larry Walker, the National League's 1998 batting champion, hails from where?

174. Where has Walker played major league baseball?

175. Babe Ruth pounded his first professional home run in what Canadian city?

176. Ned Hanlan was world single-sculls rowing champion from 1880 to 1884. Where did he start his career?

177. Where did the sport of lacrosse get its start?

178. Lacrosse in early Canada was centred in what two regions?

179. Six-day bike races were popular diversions during the Depression. The world's best racer was Torchy Peden. From what Western city did he hail?

180. Where was the game that we know as North American football popularized?

181. What team won the first Grey Cup in 1909?

182. What was the first Western city to win the Grey Cup?

183. The origins of hockey are somewhat obscure. Which three Canadian cities each claim to be the site of the first hockey match?

184. What teams battled for the first Stanley Cup in 1893?

185. Which team won the Stanley Cup in 1903, 1904, 1905, 1909, 1911, 1920, 1921, and 1923 with largely the same players?

186. What team won the first NHL championship in the league's first season of play (1917-18)?

187. The birth of Gordie Howe put *what* small Saskatchewan community "on the map"?

188. Don Cherry and his views on hockey are widespread due to his frequent appearances on Canadian radio and television. Where did he play NHL hockey?

189. What percent of NHL players are Canadian-born?

190. Canada produces the majority of NHL players. What country produces the second most – Russia, Sweden, the United States or Czech Republic?

191. Ontario is the most productive training ground for NHL players. What percent of all current NHL players hail from Ontario?

192. Ontario ranks fifth amongst the provinces in the number of NHL players produced per provincial capita at 14.6. What province produces the most NHL players per capita?

193. What city has produced the most active NHL players?

194. What city currently produces the most NHL players per capita?

195. Wayne Gretzky was born and first started scoring goals in what Ontario town?

196. Hockey's famous "Kraut Line" was part of what NHL team?

197. The "Production Line" was part of what team?

198. What team boasted the "Punch Line"?

199. What NHL team was first to start a woman player?

200. Name the father and son Formula One race-car drivers?

201. The first European people to create a settlement in what is now Canada were neither English nor French. Who were they and where did they come from?

202. Where was North America's first non-Aboriginal child born?

203. Thousands of families migrated to Canada in the mid-1800's because of the failure of *what* crop?

204. The Irish were quarantined on *what* island in the lower St. Lawrence before their entry onto the mainland?

205. What great celebration did the Irish initiate in North America?

206. French-speaking Catholics were expelled from their colonial homes in New Brunswick, Nova Scotia, and P.E.I. in the middle of the eighteenth century. Many of these Acadians – or "Cajuns" – fled to the United States. What American state do most Cajuns now call home?

207. From what country did the first great wave of settlers into Ontario come?

208. Where was the first Chinese community in Canada?

209. Ivan Pylpypiv and Vasyl Eleneak were the first immigrants to Canada from what country?

210. Where is the Canadian National Ukrainian festival held every August?

211. 396 Sikh immigrants from India aboard the *Komagata Maru* were refused entry into Canada at *what* port in 1914?

212. Most archeologists and anthropologists agree that Aboriginal ancestors crossed over to North America from *what* continent?

213. Between 1820 and 1970 over 37 million people immigrated to Canada and the United States from what continent?
A. Africa
B. Asia
C. South America
D. Europe

214. The largest number of Canadian immigrants has come from which area of the world since 1983?
A. Africa
B. Europe
C. Asia
D. South America

215. What is the current life expectancy rate for Canadian newborn girls?
A. 81
B. 85
C. 78
D. 91

216. And for newborn boys?
A. 80
B. 75
C. 70
D. 65

217. What were the life expectancy rates in 1921?

218. How many children were born in Canada's baby boom of 1946-66?
A. 910,000
B. 8 million
C. 2 million
D. 12 million

219. Indian reservations in the United States are called *what* in Canada?

220. There are over 2,000 reserves in Canada in every province but one. Name this province.

221. Do more Aboriginal peoples live on reserves or outside reserves in cities and towns?

222. Which province or territory has the largest number of Aboriginal peoples?

223. What reserve is noted for its high steel workers who have worked on the skyscrapers of New York, Montreal, and other places in eastern North America?

224. In the first half of the nineteenth century, an entire group of Aboriginal peoples was wiped out by war and disease. Who were these people?

225. A 300 year-old, 50 m tall golden Sitka held sacred by the local Haida population was destroyed by a disturbed transient in June, 1997. On what island did the tree stand?

226. Totemism is a widespread belief in kinship between people, animals and plants. What part of North America is most associated with this ancient form of worship?

227. How do Aboriginal people express their totemism?

228. Tecumseh, a Shawnee from U.S. territory, fled north to Canada to avoid persecution by U.S. forces. He was made a general by General Isaac Brock and assumed military leadership of those who escaped to Canada with him. In the War of 1812, his forces captured a large fort in the United States. Name this fort.

229. Tecumseh was killed in the War of 1812 during the Battle of Moraviantown. Where was he buried?

230. Sir Isaac Brock is considered a hero of the War of 1812. He ordered the capture of Michilimackinac and engaged in attacks on Detroit and Amhertsburg. A coin was struck in his honour, and a statue erected in his memory near the scene of his last battle. Where was the statue erected?

231. Laura Secord is a heroine of the War of 1812. What did she do and where?

232. What animal unwittingly aided Laura Secord in deceiving the enemy?

233. The Rebellion of 1837 was led by two men: William Lyon Mackenzie and Louis-Joseph Papineau. What two geographic regions did they represent?

234. Where was Louis Riel hanged?

235. How many Canadian men and women lost their lives in World War I?
   A. 41,000
   B. 51,000
   C. 61,000
   D. 71,000

236. Who is the most decorated Canadian war hero and where did he come from?

237. Where did he die?

238. Who was his famous military friend and business partner who bore the same initials, B.B.?

239. Name the famous Canadian World War II ace who died flying for the Israeli Air Force.

240. Where in Canada was the World War II spy camp, "Camp X"?

241. The Germans "occupied" what area in Canada during World War II?

242. Canada has been one of the busiest peacekeepers in the United Nations. Where did Canadians first play a peacekeeping role?

243. One Canadian politician in particular took an active role in peacekeeping and was awarded the Nobel Peace Prize in 1957. Name this politician.

244. Port Royal, the first permanent and still existing European settlement in Canada, was established by the Sieur de Monts and Samuel de Champlain in 1605, abandoned in 1607 and reestablished in 1610. In what present-day province is Port Royal located?

245. Which part of Canada did Captain James Cook explore in 1778?

246. What explorer drew maps of the coast of what is now B.C. in 1793-94?

247. Many early maps of the Canadian West were based on surveys by a British explorer who has been described as "the greatest land geographer who ever lived." Name the explorer.

248. Pierre Berton called him the greatest land explorer of his day. He was born in 1858 and died in 1957. He worked at various earth science activities all his life and retired as president from the Kirkland Lake Gold Mining Company at age 96. His many discoveries included the mineral belt on which Flin Flon, Manitoba, is located and the gold deposit at Kirkland Lake, Ontario. But it is for his 1883 discovery of dinosaur bones in the Red Deer Valley that this person is best known. Name this discoverer.

249. Which one of the following four European countries did *not* lay claim to and colonize large parts of North America?
  A. Denmark
  B. France
  C. Spain
  D. Sweden

250. In what year did John Cabot (Giovanni Caboto) first claim parts of eastern Canada for the King of England?
  A. 1447
  B. 1497
  C. 1547
  D. 1597

251. British explorer Martin Frobisher made an erroneous discovery while exploring Baffin Island in 1576. What was it?

252. The first European explorer to reach the Pacific after crossing Canada has a great river named after him. This river, however, does not flow to the Pacific. Name the explorer and thereby the river.

253. Captain Joseph Bernier was one of Canada's great marine explorers. Bernier commanded 105 ships during his lengthy career and crossed the Atlantic 269 times. However, it is for one 1909 event that he is best remembered. What was it?

254. Donald Alexander Smith, Lord Strathcona, drove in the last spike of the CPR – *where*?

255. Three important bodies of water in eastern North America were named after which early English explorer? (Hint: the water bodies include a river, a bay, and a strait.)

256. In 1670 the English Crown gave what company an area described as "all the land drained by the rivers and their tributaries flowing into Hudson Bay and James Bay"?

257. The government of Canada bought Rupert's Land from what company in 1869?

258. The goal of many early Arctic explorers was a "North West Passage"— to *where?*

259. What's the longest distance that a message sent by a Canadian has travelled?

260. Who sent it?

261. Who was the first Canadian in space?

262. Who was the first Canadian woman in space?

263. Who was the first Canadian woman to work on the International Space Station?

People

Answers

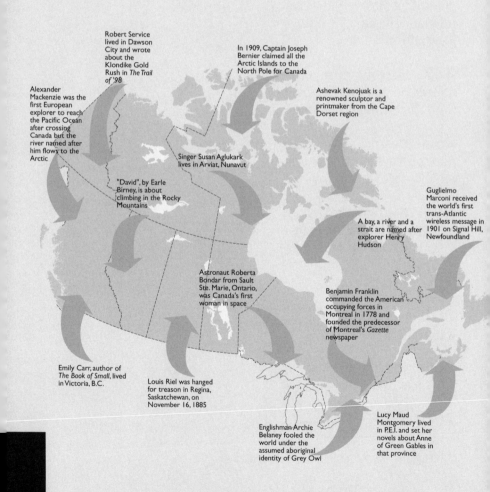

Robert Service lived in Dawson City and wrote about the Klondike Gold Rush in *The Trail of '98*

In 1909, Captain Joseph Bernier claimed all the Arctic Islands to the North Pole for Canada

Alexander Mackenzie was the first European explorer to reach the Pacific Ocean after crossing Canada but the river named after him flows to the Arctic

Ashevak Kenojuak is a renowned sculptor and printmaker from the Cape Dorset region

Singer Susan Aglukark lives in Arviat, Nunavut

"David", by Earle Birney, is about climbing in the Rocky Mountains

Guglielmo Marconi received the world's first trans-Atlantic wireless message in 1901 on Signal Hill, Newfoundland

A bay, a river and a strait are named after explorer Henry Hudson

Astronaut Roberta Bondar from Sault Ste. Marie, Ontario, was Canada's first woman in space

Benjamin Franklin commanded the American occupying forces in Montreal in 1778 and founded the predecessor of Montreal's *Gazette* newspaper

Emily Carr, author of *The Book of Small*, lived in Victoria, B.C.

Louis Riel was hanged for treason in Regina, Saskatchewan, on November 16, 1885

Englishman Archie Belaney fooled the world under the assumed aboriginal identity of Grey Owl

Lucy Maud Montgomery lived in P.E.I. and set her novels about Anne of Green Gables in that province

1. The Glenbow Museum, located in downtown Calgary, holds one of Canada's most extensive collections of Western Canadian art, photography and historical artifacts.

2. In the Confederation Gallery in Charlottetown.

3. Robert Harris, who was born in Charlottetown.

4. The McMichael Gallery, which houses the McMichael Canadian Art Collection, is located in Kleinburg, Ontario, a few dozen kilometres north of Toronto.

5. The Canadian wilderness, especially regions around Georgian Bay, Algonquin Park, Algoma, and the north shore of Lake Superior.

6. The group was formed in 1920 and consisted of Arthur Lismer, F.H. Varley, J.E.H. MacDonald, and Lawren Harris plus A.Y. Jackson, Franklin Carmichael and Franz Johnston.

7. Tom Thomson, who died in 1917, three years before the Group of Seven were officially formed.

8. He drowned in Canoe Lake in Ontario's Algonquin Park. His death is considered mysterious because Thomson was an expert in wilderness survival and an accomplished canoeist. He drowned in only a few feet of water, his overturned canoe nearby.

9. St.-Jean-Port-Joli, east of Quebec City on the south shore of the St. Lawrence River.

10. Victoria, B.C., but she travelled and painted in most areas of Vancouver Island. She detailed her life story in, The Book of Small.

11. Tofino.

12. Colville taught at Mount Allison University in Sackville, New Brunswick, from 1946 to 1963. His style of painting, now known as High Realism, influenced Christopher Pratt, who was perhaps his most famous student.

13. He was born and raised in Newfoundland, studied engineering at Memorial University in St. John's, and art at Mt. Allison and also in Scotland.

14. Ken Danby, who lives near Guelph, Ontario.

15. Robert Bateman.

16. The Prairies.

17. *Many of Kane's subjects were located in Western Canada, thanks to the patronage of Sir George Simpson who ran the Hudson's Bay Company. Simpson gave Kane a pass, which allowed travelling and boarding privileges with the Hudson's Bay Company.*

18. *Krieghoff set much of his work in Quebec and focused on the habitants of the mid-nineteenth century. He was born in Amsterdam but married a French-Canadian woman after a period of service in the American army.*

19. *Leduc was born and lived as an artist in the shadow of Mont Saint-Hilaire on the south shore of the St. Lawrence River. He lived for ninety years, and spent much of his professional life decorating churches and also producing many highly-acclaimed paintings. Many consider him the father of Quebec art.*

20. *Red Deer, Alberta.*

21. *Ashevak Kenojuak is an Inuit sculptor and printmaker. She was born in the Cape Dorset area of the Canadian Arctic.*

22. *The massive cedar sculpture is located at the University of B.C.'s Museum of Anthropology.*

23. *The Art Gallery of Ontario in Toronto holds the world's most comprehensive collection of Moore's work.*

24. *Louis Hémon, born in Brest, France, 1880.* Maria Chapdelaine *has been read by generations of Canadians, has sold millions of copies in various languages worldwide, and has been made into several movies.*

25. *Louis Hémon had been living in Canada for less than two years when he was struck by a CPR train near Chapleau in Northern Ontario.*

26. *It was published by L.C. Page of Boston. The subsequent royalties allowed L.M. Montgomery to give up teaching so she could write full time.*

27. *The majority were written in Ontario. Only* Anne of Green Gables *was written while Montgomery still lived in Cavendish, P.E.I.*

28. *World War I.*

29. *McCrae was born in Guelph, Ontario, in 1872, and received his medical education at the University of Toronto. He was a medical officer during World War I and died of pneumonia in Boulogne, France, in 1918.*

30. *Toronto. He studied and taught at the University of Toronto and also taught at York University.*

31. *The poem states explicitly that the mountains are the Rockies, and the reader is led to think of the highest, most rugged Canadian peaks because of references to geological time, fossils, and glacial features.*

32. *The University of B.C. in Vancouver, where he taught for twenty years.*

33. *Wingham. Munro's collection of stories,* The Love of a Good Woman, *was awarded the 1998 National Book Award.*

34. *Ghana, West Africa, where she once lived with her family.*

35. *Manawaka.*

36. *Irving Layton was born in Romania in 1912.*

37. *He taught English at Sir George Williams University in Montreal, which later became Concordia University. Oddly, Layton's first degree was a B.Sc. in agriculture.*

38. *The upper middle-class Montreal suburb of Westmount.*

39. *He was born in Lachine, Quebec, a working-class suburb of Montreal. His family moved to downtown Montreal before leaving for Chicago when Bellow was nine years old.*

40. *Thomas Chandler Haliburton, who lived in Windsor, Nova Scotia, before taking up residence in England at age sixty.*

41. *Glace Bay, on Cape Breton Island.*

42. *He taught school at Lower Canada College from 1935-1945 and was later on the faculty at McGill University from 1951 until his death in 1990.*

43. *The author of such classics as* A Whale for the Killing, The Boat Who Wouldn't Float, *and* The Dog Who Wouldn't Be *was born in Belleville, Ontario, but grew up in Saskatoon. He lived at various times in Toronto, Port Hope, Ontario, the Northwest Territories, the Magdalen Islands, various parts of Newfoundland and Cape Breton. Mowat also served in Europe during World War II.*

44. *Lost in the Barrens is set in the inland tundra of the Northwest Territories.*

45. Siber – *the United States government declared Farley Mowat persona non grata and banned him from the U.S. for almost 30 years.*

46. *Pierre Berton was born in Whitehorse and grew up in Dawson. His mother was a teacher and novelist noted for her account of Yukon life in,* I Married the Klondike. *Berton now lives in Ontario.*

47. *Kogawa and her family were moved from Vancouver to Slocan, near Nelson, B.C.*

48. *Weyburn, Saskatchewan, but Mitchell lived most of his life in Alberta.*

49. *Kinsella's first three novels were about life on a Cree reserve near Edmonton.* Shoeless Joe, *on the other hand, focused on baseball. Kinsella's laconic, humorous, narrative style is universally in evidence throughout his work. He continues to write about both reserve life and baseball.*

50. Field of Dreams. *Both the novel and film are set in Iowa.*

51. *Grand Falls, Newfoundland.*

52. *Both are set in Newfoundland paper mill towns, an environment Pinsent knows well.*

53. *In Davies' hometown of Thamesville, Ontario, just east of Chatham.*

54. *Davies attended Queen's University in Kingston, Ontario, where he began his career as a novelist.*

55. *Salterton.*

56. *Tom Patterson founded the Festival in 1953. The first production featured Alec Guinness in* Richard III, *and was held in a large tent on the site of the present Festival Theatre.*

57. *Timothy Findley, who was raised in Rosedale, just north of downtown Toronto.*

58. *Morley Callaghan.*

59. *Ernest Hemingway.*

60. *Hugh Garner – the novel parallels Garner's own youthful working-class experiences.*

61. *Margaret Atwood.*

62. *Atwood's novel is set at the Royal Ontario Museum in Toronto.*

63. The English Patient.

64. *He was born in Ceylon, which is now Sri Lanka.*

65. *Montreal – Anne Carson, poet, essayist, and painter of volcanoes, also lives there and teaches classics at McGill University and the University of California, Berkeley. Her novel in verse,* Autobiography of Red, *was nominated for the 1998 National Book Award.*

66. *The Prairie region.*

67. *Esterhazy, Saskatchewan.*

68. *Shellbrooke, Saskatchewan.*

69. *The weather: droughts, snow and dust.*

70. *Robert Service. He lived in Dawson City where he worked as a teller for the Canadian Bank of Commerce.*

71. *Robert Service wrote about the Chilkoot Pass which, in 1897-98, was the only route prospector-miners could take from the Pacific coast to the Yukon gold fields.*

72. *Germany.*

73. *Moore was born and lived until his mid-twenties in Ireland.*

His Governor General's Award winning novel, The Luck of Ginger Coffey, *was set in Montreal.*

74. *Montreal, where Scott became Dean of Law at McGill University. In 1977, he won a Governor General's Award for Non-Fiction for his legal treatise,* Essays on the Constitution.

75. *He was born in Quebec City and, except for fulfilling a Rhodes Scholarship at Oxford, lived his whole life in Canada. He won two other Governor General's Awards, one for his translations in* Poems of French Canada, *and the other for* Collected Poems of F.R. Scott.

76. *Toronto.*

77. *Maillet was born and raised in Buctouche, a francophone region of New Brunswick. Her novel,* Pélagie-la-charrette, *is the story of a woman who leads a group of Acadians back to their homeland after the expulsion of 1755.*

78. *Grey Owl.*

79. *Riding Mountain National Park in Manitoba and Prince Albert National Park in Saskatchewan.*

PEOPLE

80. *Kamouraska, a village on the south shore of the St. Lawrence River, a little over 100 km downstream from Quebec City.*

81. *Manitoulin Island in Lake Huron.*

82. *St. Boniface, Manitoba.*

83. *Western Bay, Newfoundland. Pratt's father was a clergyman and his grandfather a sea captain. Pratt began an academic career as a psychologist but changed direction and taught English at the University of Toronto for 32 years.*

84. *Tish was a prominent literary magazine published in Vancouver from 1961-69. Over its eight year run, Tish had many notable Canadian writers as editors, including George Bowering, Fred Wah, Frank Davey, and Daphne Marlatt.*

85. *Roch Carrier wrote "The Hockey Sweater," which was made into a short animated film by the National Film Board.*

86. *The Ouimetoscope was founded in Montreal in 1906. It was, in fact, the first movie house in North America.*

87. *Kamloops.*

88. *Donald Sutherland, who was born in Saint John, New Brunswick, and was raised in Bridgewater, on Nova Scotia's South Shore.*

89. *Bethune served on the Loyalist side in the Spanish Civil War in 1936 at which time he organized the first mobile blood-transfusion service. He later fought with the Chinese Communists in their overthrow of the Nationalist forces. During this revolution, Bethune created the world's first mobile medical service. Only well after his death in China did the world become aware that the People's Republic considered Bethune a hero.*

90. *Gravenhurst, Ontario – he trained as a doctor at the University of Toronto, and practiced medicine in Detroit and at the Royal Victoria Hospital in Montreal.*

91. *Kensington Market – The King of Kensington reigned on CBC television from 1975-80.*

92. *Chicago.*

93. *Mack Sennett, who popularized the cream-pie-in-the-face gag and whose studio produced The Keystone Kops, was from Danville in the Eastern Townships of Quebec.*

94. *Winnie is short for Winnipeg. The real Winnie lived in a London zoo but was brought from Canada during World War I by a Winnipeg serviceman.*

95.
Mary Pickford, Coquette (1929), b Toronto
Norma Shearer, The Divorcee (1930), b Montreal
Marie Dressler, Min and Bill (1931), b Cobourg, Ont.

96. *Douglas Shearer, brother of Norma Shearer, won 12 Academy awards in the field of Sound.*

97. *William Notman, assisted by his sons, was a Montrealer who set up photo studios in Montreal, Toronto, Halifax, Ottawa and a number of American cities. The Notman Collection of photos consists of over 400,000 negatives that form a detailed record of late nineteenth- and early twentieth-century Canada.*

98. *Few would argue that this honour belongs to anyone but Yousuf Karsh who hails from Armenia but set up his studio in Ottawa. His black-and-white photographs of famous people hang in collections, both private and public, worldwide. Perhaps his real genius was capturing the facial expressions of his subjects,* such as the glower on Winston Churchill's face that Karsh elicited by yanking Churchill's omnipresent cigar from his mouth just before snapping the picture.

99. *Benjamin Franklin. He founded* La Gazette Littéraire *in 1778, which was published in both French and English, and was the precursor to* The Montreal Gazette.

100. *Toronto.*

101. *Kingston.*

102. *Montreal.*

103. *An American named William Van Horne was given what Sir John A. MacDonald's critics described as an impossible task. Van Horne renounced his American citizenship and became a Canadian upon completion of the railroad, saying: "To have built that road would have made a Canadian out of the German emperor."*

104. *Bermuda (although he was probably just trying to avoid Saint John winters).*

105. *Yarmouth, Nova Scotia. A patron of the arts and education, Killam provided funds which helped establish the Canada Council.*

PEOPLE

106. *Welsford, in Nova Scotia's scenic Annapolis Valley.*

107. *Desjardins set up the first credit union, or caisse populaire, in Lévis, just across the St. Lawrence River from Quebec City. He established more than 200 credit unions in Quebec. Today, Caisses populaires Desjardins is the largest organization of its type in Canada. There are presently more than 70 million credit union members in North America.*

108. *Knowlton, now Lac Brome, Quebec. The weekly paper was called the* **Eastern Townships Advertiser,** *which Black acquired in the late 1960's.*

109. *Halifax, Nova Scotia.*

110. *County Antrim, Ireland.*

111. *The first Eaton's store was founded in 1869 and stood at the corner of Yonge and Queen Streets in Toronto. Dozens of Eaton's stores opened in what became one of the largest family owned retail companies in the world.*

112. *Halifax.*

113. *The Marquis – a quick-ripening, wind-resistant, and high-yielding spring wheat perfect for the Canadian West. Saunders held a Ph.D. in chemistry and also ran a music school. Fortunately for Western farmers, Saunders' father coerced him to join the Experimental Farms' search for a new strain of wheat – and the Marquis was born. By 1920, Marquis represented over 90% of all wheat grown in Western Canada.*

114. *University of Toronto.*

115. *In a Toronto laboratory at the Hospital for Sick Children by Dr. Frederick Tisdall in 1931. When Dr. Tisdall sold the idea for marketing the product in the United States, he ensured that royalties would go to medical research at "Sick Kids."*

116. *University of Toronto, home to Innis College. There must be few historians trained in Canada who have not referred to one of the nine volumes of history written by Innis during his less-than-sixty-year life.*

117. *Emily Howard Stowe, born in Norwich, Upper Canada, 1831. She practiced medicine in Toronto.*

118. *Dr. Stowe was forced to go to medical school in New York because no Canadian medical college would accept her. She* graduated from the New York Medical College for Women and received her licence to practice medicine in 1880.

119. Bell was a Scot who lived in both the United States and Canada. He said he developed the idea in Brantford, Ontario, in 1874. However, he actually built the first telephone in Boston the following year.

120. Pilot J.A.D. McCurdy lifted the plane off the frozen surface of Bras D'Or Lake near Baddeck, Cape Breton, in 1909.

121. Once again, Bras D'Or Lake in Nova Scotia, where the hydrofoil reached a speed of 114.04 km/hr.

122. P.E.I. and New Brunswick. This was the first permanent linking of P.E.I. to the mainland. It was probably just about this time that the first politician promised a fixed transportation link between the provinces. Many more politicians would make similar promises in the next century and a half before the $1 billion Confederation Bridge finally linked the provinces in 1997.

123. St. John's, Newfoundland, on December 2, 1901. The message was sent from Poldhu, England.

124. Valentia, Ireland, and Trinity Bay, Newfoundland. The cable was completed in 1858.

125. Fessenden was from the tiny hamlet of East Bolton in Quebec's Eastern Townships, a few kilometres from the American border. Later in 1906, he sent a message some 15 km from a ship in the Atlantic to Massachusetts in preparation for a broadcast across the ocean to Scotland. The preparation was unnecessary because the initial transmission was heard in Europe, and the Scots responded to it.

126. Toronto in 1925, by Ted Rogers, who invented the tube that allowed battery-less transmission. Rogers also started CFRB in Toronto, the first radio station to use this technology.

127. Pointe Claire, Quebec, a suburb of Montreal on the shores of Lac St. Louis.

128. This world-class aircraft was developed and built by A.V. Roe Aircraft at Malton, Ontario, between 1953-59. The Diefenbaker administration cancelled the project in 1959.

129. Robert Noorduyn built the Norseman in Montreal in 1935, using facilities that are now part of Canadair of the Bombardier empire.

PEOPLE

130. *The variable pitch propeller was invented by Wallace Turnbull in his home workshop in Rothesay, New Brunswick.*

131. *The Canadian Snowbirds are the only aeronautic flight demonstration team in existence outside Europe. The nine plane team is a familiar sight, performing at over 65 airshows in Canada and the United States every year.*

132. *In Buxton, now called North Buxton, just west of Chatham, Ontario.*

133. *A brave visionary named William King, often overlooked by Canadian historians. Through an inheritance from his father-in-law, King found himself the owner of 15 Louisiana slaves. King wanted to free the slaves and took them north to Canada. In November of 1849, the Presbyterian Church and local citizens helped King acquire 9,000 acres for the people he removed from slavery. Within three years, the residents had built a brickyard, sawmill, and gristmill. They also built a school that readied its students for university. The school was so good that white residents sought and received permission for their children to attend it.*

134. *Montreal.*

135. *Lévesque's family lived in New Carlisle on the Gaspé peninsula of eastern Quebec. His mother gave birth to René in the closest hospital, which happened to be across the border in Campbellton, New Brunswick.*

136. *Africa.*

137. *Lake Superior.*

138. *Michigan.*

139. *Gordon Lightfoot.*

140. *The Rockies and the St. Lawrence, respectively.*

141. *100th meridian, which runs through Manitoba.*

142. *Kingston.*

143. *Dan Ackroyd.*

144. *Sir Wilfrid Laurier University in Waterloo, Ontario.*

145. *Liona Boyd, who was born in London, England, in 1949, and grew up in Toronto.*

146. *Susan Aglukark hails from Arviat in Nunavut. She is a singer of Inuit and pop music.*

147. *k.d. lang.*

148. *Natashquan –
Vigneault's classic song "Mon
Pays" was written for the 1964
National Film Board film,*
La Neige a fondu sur la
Manicouagan. *The song begins
with the phrase, "Mon pays c'est
n'est pas un pays, c'est l'hiver"
(My country is not a country, it's
winter).*

149. *Montreal. In 1975, he
was made Companion of the
Order of Canada.*

150. *The Guess Who* origi-
*nated in Winnipeg in 1965 and
quickly gained long-lasting international success.*

151. *Reach the #1 spot on the
American hit parade charts.*

152. *The title is "Running
Back to Saskatoon."*

153. *He first appeared with
the Toronto Symphony Orchestra.*

154. *The piano.*

155. *York University, Toronto.
Peterson grew up in Montreal
and quickly became famous when
he debuted as a twenty-four-year-
old pianist at Carnegie Hall in
New York.*

156. *London, Ontario.*

157. *Speedboat racing.*

158. *Toronto, where they were
trained as singers at St. Michael's
College.*

159. *Anne Murray has won
twenty-five Junos over a career
which dates back to the 1960's.
She was born in Springhill, Nova
Scotia, but calls Thornhill,
Ontario, home. Gordon Lightfoot
has won the second most Junos
with 17. Lightfoot hails from
Orillia, Ontario.*

160. *He was born in Cabri,
Saskatchewan, in 1921.*

161. *The legendary concert
that succeeded without
Lighthouse was Woodstock, held
on a site 96 km southwest of
Woodstock, New York.*

162. *Almonte, Ontario, just
west of Ottawa. The McGill grad-
uate invented basketball at a
YMCA in Springfield,
Massachusetts, in December,
1891.*

163. *Yellowknife – the annual
event involves athletes from the
Yukon, Northwest Territories,
Nunavut, and Alaska competing
in traditional Inuit and First
Nations games.*

PEOPLE

164. *Just outside Thunder Bay, Ontario. A most impressive memorial containing rock from every Canadian province marks the spot along the Trans-Canada Highway.*

165. *St.-Cyprien-de-Napierville, near Montreal. Cyr was a Montreal policeman but left his profession to become a travelling strongman. His most famous feat was performed in Boston in 1895, where he lifted 1,967 kg on his back. The weight consisted of 18 large men.*

166. *Lake Ontario. In 1954, when only sixteen years old, Toronto-born Bell swam from Youngstown, New York, to Toronto.*

167. *The man Muhammed Ali called "the toughest man I have ever fought" was born in Toronto.*

168. *This is a bit of a trick question. The first Canadian to win Olympic Gold was George Orton in the 2,500 m steeplechase at the 1900 Paris Olympics. However, Orton competed for the United States because Canada sent no team that year. The first Canadian Gold medals were won by four individuals at the next Olympics in St. Louis in 1904: Étienne Desmarteau for weight throw; George Lyon in golf; the*

Winnipeg Shamrocks for lacrosse; and the Galt, Ontario, squad in soccer.

169. *Brooks was a Torontonian; her three world titles were won in Saint John, New Brunswick.*

170. *Robinson played his only minor league season for the 1946 class AAA Montreal Royals where he led the team to the league championship.*

171. *Jenkins, winner of 284 games and credited with striking out a total of 3,192 batters, hails from Chatham, Ontario.*

172. *Jenkins broke in with the Philadelphia Phillies but spent most of his career starring for the Chicago Cubs. In the latter part of his career he played for the Texas Rangers and the Boston Red Sox before retiring with the Cubs.*

173. *Walker calls Maple Ridge, B.C., home.*

174. *Walker broke in with the Montreal Expos before moving on to the Colorado Rockies where he won the National League's Most Valuable Player Award in 1997.*

175. *Ruth hit his first pro home run at Hanlan's Point Stadium on the Toronto Islands* in 1917. Ruth also pitched in the game, surrendering only one hit in a 9-0 win.

176. Hanlan started rowing in Lake Ontario at Toronto.

177. This highly-ritualized athletic event was widely played throughout eastern North America by many different Aboriginal peoples, well before the first Europeans came to North America.

178. Caughnawaga, near Montreal, and Akwesasne, near Cornwall, Ontario. Both are now Indian reserves.

179. Victoria, B.C.

180. In 1874, Harvard and McGill played two exhibition football matches in Montreal, one with Harvard's slow rugby pass-back rules and one that allowed for greater passing latitude. Harvard liked the faster, hands-on McGill game and brought it back to the U.S. from Montreal where it spread to other universities and eventually evolved into American football as we know it today.

181. The University of Toronto Blues won the Grey Cup for the first three years of competition: 1909, 1910, and 1911. At that time amateur teams competed for the cup, too.

182. The Winnipeg Blue Bombers won the Grey Cup in 1935.

183. Dartmouth, Nova Scotia, Montreal, and Kingston, Ontario, all claim to have hosted the first game. The original games were forms of shinty, or hurley, which are acknowledged to be the forerunners of hockey. Dartmouth proponents claim the first games in their area were played by Aboriginal peoples who lived in the area more than a century before the date given by either the Montreal or Kingston proponents. By Confederation in 1867, ice hockey, played on skates and similar to the game we know today, was quite widespread. By 1875, the first indoor games were being played with a puck on a rink with the same dimensions as rinks today.

184. The Montreal AAA defeated Lord Stanley's favourite, the Ottawa Senators.

185. The Ottawa Senators – they had a talented group of players known as the 'Silver Seven.'

186. The Toronto Arena defeated the Montreal Canadiens for the league title, and then went on to defeat the Vancouver Millionaires for the Stanley Cup.

PEOPLE

187. *Floral – Howe was born there in 1928.*

188. *Boston, where he dressed for one NHL game. He later coached the Boston Bruins and Colorado Rockies for a total of six seasons.*

189. *60% – just over 400 of the 670 1998-99 season players were born in Canada.*

190. *For the 1998-99 season, Americans made up 15% of the players, Russia 7%, Czech Republic 6%, and Sweden 5%.*

191. *Almost 25% of all players, or about 40% of all Canadian players hail from Ontario.*

192. *Saskatchewan with 34 players from a population of 1,019,632, which works out to a rate of 33.3 players per million. Alberta is second with 21.5 and Manitoba is third with 15.8.*

193. *Montreal, with a total of 29 players. This works out to 2.9 players per 100,000 citizens.*

194. *Belleville ranks first with 10.8 players per 100,000 citizens. Kingston, a few kilometres east of Bellevile on Highway 401, comes second with 10.4 per 100,000.*

195. *Brantford, in 1961.*

196. *Milt Schmidt, Bobby Bauer, and Woody Dumart (all Canadians of German descent) played together for the Boston Bruins from 1939-41.*

197. *Ted Lindsay, Sid Abel, and Gordie Howe – known simply as Lindsay, Abel and Howe – played for the Detroit Red Wings.*

198. *Maurice Richard, Toe Blake, and Elmer Lach played for the Montreal Canadiens.*

199. *Manon Rheaume started an exhibition game in goal for the Tampa Bay Lightning on September 23, 1992. She left at the end of the first period in a 2-2 tie with the St. Louis Blues. Rheaume played major junior hockey in Trois-Rivières, Quebec, and was the leading goalie for the Canadian women's hockey team.*

200. *The late Gilles Villeneuve and his son, Jacques, are Canada's most famous Formula One racers. Gilles was raised in Berthierville on the north shore of the St. Lawrence River near Trois-Rivierès in Quebec. Jacques was born in St. Jean-sur-Richelieu near Montreal and was raised in Monaco.*

201. *The Norse from Scandinavia.*

202. *Snorri Thorfinnsson was born to Viking traders on what is now Newfoundland, in either 1003 or 1004.*

203. *The potato crop in Ireland.*

204. *Grosse-Île, just east of Quebec City. Many Irish immigrants contracted typhoid fever (ship fever) in the abominable conditions aboard the "coffin" ships that carried them from Ireland and England. In 1847 alone, 5,293 died at sea and an additional 8,072 perished on Grosse-Île. Survivors went on to live in the large Irish ghettoes of Montreal and Quebec City.*

205. *The St. Patrick's Day Parade. Montreal hosts the longest continuously running parade in the world, which attracts hundreds of thousands of spectators each year (all of whom claim to be Irish that day).*

206. *Louisiana.*

207. *The United States. Many were United Empire Loyalists who headed to Canada during and after the American Revolution.*

208. *Barkerville, B.C., established in the mid-nineteenth century.*

209. *Ukraine, in 1891.*

210. *Dauphin, Manitoba.*

211. *Vancouver – the Sikhs were refused entry after waiting two months aboard ship. Their rejection preceded a year of race riots and immigration law disputes.*

212. *Asia – they crossed over a land/ice bridge that covered the Bering Strait during the Wisconsin glacial period.*

213. *D. Europe*

214. *C. Asia*

215. *A. 81*

216. *B. 75*

217. *Newborn boys had a life expectancy of 59 years, girls a life expectancy of 61 years.*

218. *B. 8 million*

219. *Reserves.*

220. *Newfoundland.*

221. *More than half of Canada's Aboriginal peoples live in towns and cities. Only 19% live on reserves.*

222. *Ontario leads the way with over 252,000 people, followed* by British Columbia with 174,000. Alberta has 159,000 people while Quebec is fourth with over 140,000 Aboriginal peoples.

223. The Mohawks of Caughnawaga on the south shore of the St. Lawrence River near Montreal. Thirty-five of the workers killed in the 1907 Quebec Bridge accident came from Caughnawaga.

224. The Beothuks. The last Beothuk died in 1827 and her death marked the elimination of the last of a tribe that numbered 50,000 at the time of European contact. The Beothuks perished in battle with the English and Mi'kmaq, and also from diseases like smallpox and tuberculosis. They lived in Newfoundland.

225. Graham Island, B.C.

226. The Pacific Northwest – Coastal peoples have maintained totemistic beliefs for centuries.

227. With totemistic symbols carved and/or painted on posts called totem poles. They are erected outside dwellings or as memorials to the deceased.

228. Fort Detroit, which Tecumseh captured with a force of 600 warriors against 2,000 American defenders.

229. In order that his body not be desecrated by American soldiers, Tecumseh's followers buried him in an unmarked grave somewhere in southwestern Ontario.

230. The 64 m high statue was erected at Queenston Heights, a short 11 km drive from Niagara Falls. The statue stands on the site where Brock is buried.

231. In 1813 she walked over 30 km through enemy lines to warn English commander James FitzGibbon of an impending ambush by American forces.

232. A cow and two empty milk buckets gave the mother of five a rather innocent appearance thus allowing her to walk freely through American lines.

233. Mackenzie represented Upper Canada and Papineau represented Lower Canada. Generally, Upper Canada included what is now southern and eastern Ontario. Lower Canada covered the Quebec portion of the St. Lawrence Valley.

234. Métis leader Louis Riel was executed for treason in Regina on November 16, 1885. On a morbid note, the noose used to hang him was on display until recently in Regina's RCMP museum. *Hanging those convicted of a capital crime remained a legal option in Canada well into the 1960's.*

235. *C. 61,000*

236. *William Barker won more decorations for gallantry than any other Canadian. During World War I, he shot down over 50 enemy planes. He is best known for his single-handed dog-fight with over 60 German Fokker D VII's. Barker survived the fight despite receiving a severe bullet wound in the leg and crashing behind British lines. This display of courage earned him a Victoria Cross, the Common-wealth's most prestigious award. Barker was born in Dauphin, Manitoba.*

237. *An aircraft accident at Rockcliffe in Ottawa took Barker's life in 1930. He was 35 years old.*

238. *Billy Bishop, who was from Owen Sound, Ontario. He single-handedly shot down 72 German aircraft during World War I and was the first Canadian pilot to receive the Victoria Cross.*

239. *Buzz Beurling – born in Verdun, Quebec, in 1921. He shot down 28 enemy aircraft while serving with the Royal Canadian*

Air Force in World War II. Beurling left the RCAF in 1944 and joined the Israeli Air Force. He died in 1948 while ferrying aircraft from Rome to Palestine.

240. *Winnipeg-born Sir William Stephenson, who was immortalized in the book, A Man Called Intrepid, set up the camp on the shores of Lake Ontario, near Oshawa, Ontario.*

241. *The crew of a German U-Boat landed at Martin Bay in Northern Labrador and set up a battery-powered weather station which transmitted much needed weather information to the German fleet. The station was located in such a remote site that it went undiscovered for the next thirty-seven years and was found only when one of the German crew who set-up the installation asked the Canadian government what had happened to it.*

242. *Kashmir in 1948 where Canadian officers supervised the shaky truce between India and Pakistan.*

243. *Lester B. Pearson – Canada's fourteenth prime minis-ter was also former president of the General Assembly of the United Nations.*

244. *Nova Scotia. The oldest*

settlement in North America, however, is St. Augustine, Florida, based on the founding of a French Fort there in 1564.

245. The Pacific coast. In his previous two trips to Canada, Cook participated in the siege of Louisbourg on Cape Breton, and mapped the Gaspé peninsula, the estuary of the St. Lawrence, St. John's harbour, and the Newfoundland coast.

246. Captain George Vancouver.

247. David Thompson.

248. Joseph Tyrrell. The Royal Tyrrell Museum near Drumheller, Alberta, was named in his honour. Two days after his dinosaur bone discovery, Tyrrell discovered Canada's largest deposit of bituminous coal in Drumheller.

249. D. Sweden

250. D. 1597

251. Convinced he had stumbled on a huge gold deposit, Frobisher loaded his ship with iron pyrite and raced back to England and Queen Elizabeth I. Imagine Frobisher's chagrin when experts told him his mighty fortune was nothing but a pile of fool's gold.

252. (Alexander) Mackenzie.

253. At Melville Island in July, 1909, Bernier claimed the Arctic Islands for Canada, including the territory from 60° west longitude to 191° west longitude and as far north as the pole at 90° north latitude.

254. Craigellachie, B.C. Smith also became the wealthiest man in Canada, rising from factor at various out-posts in the Hudson's Bay Company to owning and running the company. A shrewd businessman, he invested wisely in the coming railway and oversaw his investment by assuming control over construction and becoming president of the Canadian Pacific Railway. At various times Smith held seats in the House of Commons, representing Selkirk, Manitoba, and later Montreal West. He was president of the Bank of Montreal until he retired at 85. He was also Chancellor of McGill University and, until his death in 1914, Canada's High Commissioner in London, a post that he occupied with his accustomed aplomb.

255. Henry Hudson.

256. Hudson's Bay Company.

257. Hudson's Bay Company.

258. *China.*

259. *A message sent in 1969 travelled 384,400 km – to the moon.*

260. *The loony-sized disc contained 73 messages and was left on the moon by the first lunar astronauts. Prime Minister Trudeau was the author of the Canadian message which read, in English: "Man reached out and touched the tranquil moon. May that high accomplishment allow man to rediscover the earth and there find peace."*

261. *Astronaut Marc Garneau, born in Quebec City, flew on the space shuttle Challenger in 1984.*

262. *Astronaut Roberta Bondar, who was born in Sault Ste. Marie, flew on the space shuttle Discovery in 1992.*

263. *Astronaut Julie Payette, born in Montreal, worked on a ten-day logistics and resupply mission in May, 1999.*

PEOPLE

Guide to Using Index

The first number in the sequence (1:1) refers to the page; the second number refers to the appropriate question or answer.